The Emperor's Quantum Computer

An Alternative Light-Centered Interpretation of Quanta, Superposition, Entanglement and the Computing that Arises from it

Pravir Malik, Ph.D.

ISBN-13: 978-0-9903574-8-3

To Sri Aurobindo

This book leverages an alternative interpretation of Quantum Theory explored in the *Cosmology of Light* book series, to suggest an alternative way to conceive of the fledgling field of Quantum Computing.

In *Cosmology of Light* quanta are a means of precipitation by which information codified in light moving faster than c, materializes in a reality where light travels at c. This book will adequately summarize the conceptual and mathematical framework that animates a cosmology of light and use that as a basis to suggest an alternative genre of quantum computing.

A philosophy of computation necessitates that possibilities of computation will derive from some perception of reality. Common digital computers derive from proven laws of electricity and their effect on binary-logic based devices at the physical level. Research and development at the physical level has proceeded for decades, and there is a relatively detailed and practical understanding of leveraging many physical laws. Our conception of quantum computation will similarly derive from our conception of quantum constructs such as the nature of quanta and attendant dynamics such as superposition and entanglement. In contrast to the physical layer, which 'sits on' and derives stability from deeper phenomena at the quantum-levels, the quantum-levels are only a window into antecedent layers created by light traveling at different speeds. To assume therefore that notions of computation true of the physical layer can be easily projected onto the quantum-levels and that we can therefore compute the same information as we do at the physical layer faster and in vaster quantities is questionable.

This book will explore the dynamics of superposition and entanglement from the point of view of precipitating layers of reality so set up light traveling at slower and slower speeds down to c, to in fact arrive at a different notion of quanta, of superposition, and of entanglement, that will suggest that reality at the quantum-level may be different from the view commonly held today.

The notion of superposition that is the foundation of modern day quantum computing derives from the Copenhagen Interpretation of Quantum Mechanics that theorizes that until a photon, or particle, or other quantum-level object is measured, it exists in an infinite number of superposed states that collapses into a perceivable state only through the act of measurement. But if that were the case, then there is no basis for the universe having come into being, or evolving the way it has, or producing a human species on planet earth, since there was conceivably never any measurement to have precipitated any of those outcomes in the first place.

If such outcomes could materialize, and the universe is the proof that it has, then the model on which the current notion of superposition is based, and therefore our very notion of superposition is likely incomplete. And this becomes important because further, the supposed extraordinary processing power of quantum computers is also based on this unproven or incomplete notion of superposition.

By way of a telling anecdote I was recently at a Case Institute of Technology alumni event in Palo Alto honoring Donald Knuth of 'The Art of Computer Programming' fame (Knuth, 1968). Someone from the audience made a comment about the billions of dollars big corporations such as Google, IBM, and Microsoft were pumping into Quantum Computing and asked Dr. Knuth for his opinion about quantum computing. He replied something to the effect: "Let them do what they are doing. I will continue to remain focused on classical computing."

In the Light-Centered Quantum Interpretation offered in this book the meaning of quanta, of superposition, of entanglement, amongst other quantum phenomena are revisited and another light-centric mathematics constructed. I feel justified in doing so especially since of late figures such as the sixteenth century polymath Jerome Cordoba have been positioned as being at the forefront of contributors to the mathematical foundations of quantum physics. Cordoba is afforded this status because he invented probability theory and was also amongst the first to recognize the validity of imaginary numbers, as pointed out by Michael Brooks in his book 'The Quantum Astrologer's Handbook" (Brooks, 2017). But if we have to revert to the sixteenth century at arguably the beginning of the Scientific Age to explain findings in the most recent years of the twenty-first century then something seems to be not quite right.

In the Light-Centered Quantum Interpretation, Light existing at a native state, characterized by an infinite speed, gives insight into aspects of the nature of what light must be. A projection from that native state to a state in which it travels at a finite speed, c, is suggested to have created quanta and matter as we experience it in this universe. Hence the Big Bang itself is a result of light slowing down to c, and subsequently everything that appears in the universe is already deeply entangled by virtue of having originated from a single point.

Quanta and matter therefore are of the nature of Light. The nature of light always remains the same regardless of how it projects itself so that in mathematical terms *symmetry* is maintained. Such symmetry expresses itself in a superset of the wave-particle construct inherent to the De Broglie–Bohm Pilot-Wave Interpretation of Quantum Mechanics. In this interpretation the wave and the particle exist simultaneously, so that, as seen in the famous double-slit experiment, particles or photons or other quantum-level objects are guided to where waves constructively interfere. The superset can be thought of as a block-field-wave-particle quadrality, of which wave-particle duality is an observable subset.

Further, commonly accepted quantum mechanics interpretations such as the Copenhagen and Ensemble Interpretations erect statistical and probability functions as co-equal with the nature of reality. That is, the nature of reality is taken to be statistical and probabilistic. This is justified by suggesting that while individual phenomenon are entirely probabilistic or random, in aggregate they become deterministic or predictable. But alternatively rather than such statistical aggregation, in the Light-centered interpretation of Quantum Mechanics suggested in this book, reality can be modeled as being "functional" as will be explained in greater detail, which then allows a process of *statistical disaggregation* to take place whereby apparent randomness can be seen as a bi-product of function-focused native light precipitating into more material layers that come about when light travels at c. Furthermore, Schrodinger's Wave Equation and Heisenberg's Uncertainty Principle, pillars of modern-day views of quantum mechanics can be interpreted differently to in fact reinforce the notion of reality as consisting of multiple layers of light, as will also be discussed in this book. Briefly, in such a functional-view of the universe the granularity of analyses cannot just be quantum-level particles as is currently the case, but quantum-functionality in which relationships, function, and therefore laws are implicit in the supposedly smallest phenomenon. As will be proposed, all phenomena are governed by one equation and the fullness of all the possibility of light is present in the seemingly most disaggregated, miniscule, and separate thing.

Further the very stuff of the physical, including cosmic prerequisites of space, time, gravity, and energy, can be modeled as emergent from a reality of light. There is a supra-physical process that creates the physical. To therefore approach quantum computing in a purely 'physical' manner is ab initio likely incomplete. In fact classical computing itself has a supra-physical process, such as thought-based algorithms, applied to the physical layer of bits that has made it successful. For quantum computing to be successful superposition, entanglement, and any other quantum 'weirdness' has similarly to become a repeatable process that needs to be consciously applied to, rather than 'collapsed' into, the physical layer. Either the physical layer therefore has to be such that it allows or records fullness of any such supra-physical subjectivity to be understood, or the conceptual framework of quantum computation enhanced to allow interplay with possibly multiple layers beyond just the physical.

If, as the light-based interpretation of quantum mechanics suggests, superposition and entanglement are intimately associated with the dynamics of space and time then the case can be made that they are not just accessible by going down into the quantum realms. In fact this book will highlight the place that human capacities such as will, thought, and emotion, can have in mobilizing quantum realms. If this is true then really what may be possible is a genre of human-quantum computational models.

The very basis of modern-day quantum computing that relies on infinite number of superposed quantum states, on probability, on observable measurement that brings things into reality, is bought into question in the Light-centered Interpretation discussed in this book. In fact from the point of view of the latter interpretation superposition, entanglement, and reality take on a different meaning and the infinite processing power allegedly true of quantum states, like the new clothes in Han Christian Andersen's *The Emperor's New Clothes* (Andersen, 1837) simply does not exist in the manner in which it has been conceived. The premise in Andersen's tale is that only those who are unfit, stupid, or incompetent, cannot see the clothes. Hence all the courtiers, village people, and the emperor himself pretend to be able to see these clothes for fear of being called out as unfit, stupid, or incompetent.

While MIT's Seth Lloyd leverages a quantum superposition, probability-based computational approach to the development of the universe in his book 'Programming the Universe' (Lloyd, 2007), the Light-centered Interpretation of Quantum Mechanics discussed in this book will propose an alternative view of superposition, entanglement, and computational power, founded on a quaternary-based, symmetrical, function-rich model of Light that animates all things from the invisible to the material. Such a view may give rise to a different genre of quantum computing in which the universe itself is a result of constant quantum-level computation.

The book is structured into nine sections.

Section 1 explores some fundamental concepts to do with light and quanta. Specifically, Chapter 1.1 explores the impact that light has on the experienced nature of reality. In particular the impact of light traveling at potentially different speeds will be examined for its ability to create realities based on its speed. Further the Big Bang and a pervasive seed-equation are posited as existing as a result of the slowing down of Light. Chapter 1.2 explores a theory for the creation of quanta related to the speed of light. Quanta are intimately associated with the seed-equation and acts as a doorway, as it were, into the multi-layered structure represented by the seed-equation. But time, space, and gravity can also be thought of as existing through quantization. Chapter 1.3 summarizes some possible problems with current-day quantum computing as conceived. These have to do with superposition, the basis of a physical as opposed to a supra-physical process, and the reliance on probability and statistics.

Section 2 explores the mathematical foundation for a Light-based interpretation of quantum phenomena. This section explores the mathematics required to interpret the universe as a play of functional-richness. The Light-Matrix describes the overarching structure from a ubiquitous point-instant to the vast diversity of life. The Light-Matrix also elaborates an alternative model of superposition to that held true from the Copenhagen Interpretation of Quantum Mechanics. Each layer in the matrix is then explored in more detail through the point-instant, the architectural forces, uniqueness of organizations and its emergence, and the always-available, summary, inherent-dynamics. Exploration of the dynamics of each layer also sheds insight into the phenomenon and variability of entanglement. Further, dynamics of possibility are explored through qualified determinism. Qualified determinism presents the mathematics of selecting a likely outcome from the multi-layered superposition-cum-entanglement dynamics existing at the quantum level.

Section 3 will explore the mathematics of quantization and the foundation of human-quantum computational models. Building on the mathematical foundation laid out in Section 2, this section will further elaborate some mathematics in integrating quantum-levels with human-based dynamics. Specifically this section focuses on the derivation of the Light-Space-Time Emergence equation leveraged in subsequent quantization analyses. The Light-Space-Time Emergence equation models the basis for quantization suggesting emergent reality for all phenomena from Light. Schrodinger's wave equation and Heisenberg's uncertainty principle are also interpreted from the point of view of the Light-based Interpretation of Quantum Mechanics to reinforce the notion that even when considered from these points of view the existence of multiple layers of light is feasible. The chapter on quantization of space, time, matter and gravity, models how these phenomena are related to Light and subsequently also models how these fundamentals work together to impact material reality. Finally, deductively derived mathematical operators that may be representative of manipulating quantum phenomena as per this light-based interpretation will also be explored thus suggesting some basis for human-quantum computational models.

Section 4 will explore an overview of real-time Light-Space-Time Matrix computations in the development of the universe. If the mathematics presented in the previous sections is true, then the universe becomes the result of continuous, real-time Light-Matrix based computations implicitly involving quantum-levels. In fact if this is the case, then nothing can truly change unless superposition and entanglement are leveraged. This section therefore briefly outlines how it may be that Light-Matrix based computations have created everything from the Big Bang to modern day Global Civilization.

Section 5 will explore the computation involved in the creation of the electromagnetic spectrum. The electromagnetic spectrum is a technical way to refer to light, essentially because amongst its range of properties it displays a tangible and simultaneous electric and magnetic or "electromagnetic" reality as well. So as light becomes more concrete to us or as the possibilities within it begin to emerge, one of the first forms it takes is as the electromagnetic spectrum. It must be the case that the previously surfaced properties of light – Presence, Power, Knowledge, and Harmony – emerge so as to define the very architecture of the electromagnetic spectrum. Chapter 5.1, Quantum Computations in Emergence of Electromagnetic Spectrum Logic, summarizes the emergence of the electromagnetic spectrum in terms of the underlying Light-Space-Time Emergence equation and the process of quantization that must occur to create the logic of the electromagnetic spectrum ecosystem that precipitates into the material-fabric. So we find that the four underlying properties of Light that we call Harmony, Knowledge, Power, and Presence are of the essence of the speed with which the electromagnetic spectrum moves, the wave-range within the electromagnetic spectrum, the energy-gradient within the electromagnetic spectrum, and the mass-possibilities due to the electromagnetic spectrum, respectively. Further the description – electromagnetic – seems to have captured the Power-Harmony aspects implicit in light. In reality the electro-magnetic spectrum can likely be more completely described as electro-magnetic-wavearchetype-masspotential spectrum.

Section 6 will explore the computation involved in the creation of matter. But further, we will find that layers of matter – quantum particles, which include bosons, and atoms – are also structured or emerge along the same property-lines or property-families of light. There is similarly a continuous process of computation that involves the quantum-realms and quantization to create the realities of quantum particles, including bosons, and atoms. Hence chapters, 6.1, 6.2, and 6.3, describe a process of computation by which Light emerges as quantum particles, bosons, and atoms respectively.

Section 7 will explore the computation involved in the creation of life. This section explores the quantum-level computation that causes the emergence of life through the cell, complex human attributes such as thoughts and feelings, and uniqueness of individuality. Hence Chapter 7.1 will explore the quantum-level computation in the emergence and complexification of four-foldness through the primary molecular plans of nucleic acids, proteins, lipids, and polysaccharides. Chapter 7.2 will relate key human attributes of sensations, urges, desires, wills, feelings, emotions, and thought to the continuing journey of fourfold complexification. Chapter 7.3 will relate truer individuality to the fourfold properties implicit in Light.

Section 8 will explore the computation involved in the creation of complex organization. Having traced the computations resulting in the emergence of space-time-energy-gravity, the electromagnetic spectrum, matter, life, and even human becoming as manifest in sensation, urges, feelings, and thoughts, and truer individuality, we now turn our attention to study the computation resulting in the emergence in complex organizations. Chapter 8.1 examines the emergence of mega-organizations. Chapter 8.2 examines the emergence of sustainable global civilization.

Section 9 focuses on rethinking quantum computation. Having elaborated on alternative ways to think of superposition, entanglement, and the quantum, and proposed a cohesive light-based mathematics by which quantum computation continues to compute all emergence of Light, this Section will begin to explore alternative ways in which the field of quantum computation may proceed. Chapter 9.1 summarizes four-fold complexification as possibility in Light continues to emerge, also suggesting an alternative stratum for quantum-computation - that of the quantum-level material-fabric. Chapter 9.2 will explore a philosophy of computation by contrasting the digital, qubit, and material-fabric stratums for computation, and some implications. Chapter 9.3 will explore some alternative paradigms for computation by focusing on the implicit characteristics of the material-fabric stratum.

This book is an attempt to reveal something of the vastly different and fundamentally creative quality of computing that must accompany any computation involving the quantum-levels.

To linearly project digital computing laurels manifest as increasing speeds and the ability to process vaster amounts of information, as the inevitable trajectory of quantum computing is perhaps, in the aphorism attributed to the Buddha, to look only at the finger, and to miss the moon and the sky that it is pointing to.

Pravir Malik,
San Francisco

SECTION 1: FUNDAMENTAL CONCEPTS

Chapter 1.1 explores the impact that light has on the experienced nature of reality. In particular the impact of light traveling at potentially different speeds will be examined for its ability to create realities based on its speed. Further the Big Bang and a pervasive seed-equation are posited as existing as a result of the slowing down of Light.

Chapter 1.2 explores a theory for the creation of quanta related to the speed of light. Quanta are intimately associated with the seed-equation and acts as a doorway, as it were, into the multi-layered structure represented by the seed-equation. But time, space, and gravity can also be thought of as existing through quantization.

Chapter 1.3 summarizes some possible problems with current-day quantum computing as conceived. These have to do with superposition, the basis of a physical as opposed to a supra-physical process, and the reliance on probability and statistics.

Chapter 1.1: Light and the Realities It Creates

This chapter will look at the impact light has on creating practical realities.

Light@c, Quanta, and The Reality It Creates

Everything is made from light. But how can this be. How can the dense material objects that make up our world – the stones, cars, buildings – be made from light? How can our thoughts, our emotions, our cells and bodies be made from light? How can all the animals and plants and clouds be made from light? Here is an exploration on now this may be possible.

To begin this exploration think for a minute as to how fast light is traveling. If we turn on a light switch in a room it immediately gets filled with light. If we turn on a light torch on a dark night it immediately lights up a path through the darkness. And yet the light from the sun takes about eight minutes to reach the earth – this speed through a vacuum is referred to as the physical constant 'c'. So while light is moving incredibly fast it is still not moving at the infinite speed we may think it is moving at.

Now, it is because the light is not traveling at an infinite speed that quanta and subsequently the world of matter can appear. All matter is made from atoms. So let us enter into the world of a hypothetical atom in the process of creation to explore how this might be. Imagine that some form of light originates from what will become the nucleus within an atom. By virtue of its finite speed light would take some time, however small, to travel any minute distance within the forming atom. During that fraction of time there is a build up of energy and it is this build up of energy that forms a packet or quanta, as it were, that allows matter to form. In the absence of such a build-up potential-matter would simply disperse. Hence, it is this build up on energy that variously expresses itself as quarks and leptons and bosons and subsequently atoms – the very building blocks of everything that is.

So, in this view, it is because of the finite speed of light, c, that matter is created. That is not to say that matter will be created wherever light is traveling at c, but the possibility exists, so long as other conditions are also fulfilled. But let us leave aside the other conditions and the exact process of such a creation for now.

But what else is implicit or made possible because of the finite speed of light? Well, imagine traveling on a ray of light from the sun to the earth. Imagine that you are in minute 4 of the approximately 8 minute journey. As you look back you will see that 4 minutes in the past you were at the sun. 4 minutes in the future you will be at the earth. And in the present moment you are somewhere between the sun and the earth. So this limited speed of light already creates the concept of time and specifically of the past, the present, and the future.

So, four incredibly fundamental things are created because of the finite speed of light: quanta and therefore matter, the past, the present, and the future. It could be said, therefore, that this matter-based time-bounded universe is a result of the finite speed of light.

Looking at this in another way, it is known too that in a denser medium the speed of light further slows down. Hence, light travels at some fraction of c when moving through water for instance. So by reversing the logic of such a process this may show that if light slows down from an infinite speed to some other lesser speed whether a multiple or a fraction of c, then the material reality will have to alter, in a similar way as the material reality between vacuum and water is different.

But what if light can be experienced at other speeds? What if light can travel at other finite or an infinite speed beyond the c we experience it at, which by the way is an astronomical 186,000 miles per second in a vacuum?

Let us imagine for a minute that light can travel infinitely fast. Think about a big area or volume with a light source at the center. Now since the light travels infinitely fast it will fill up the entire volume instantly. This will be true no matter how large the volume is – it could be the entire solar system, an entire galaxy, or an entire universe. So that light is going to be instantaneously present everywhere – it is going to be omnipresent.

A further thought experiment may give insight into such omnipresence by considering the night sky. It is because of the finite speed of light that the night sky has only spots of light, however many, across it. But if light were traveling infinitely fast then the night sky would be a canvas or a sea of brilliant white since the light from every corner of the universe, wherever a light source or star existed, would immediately be present. It would appear, in such a thought experiment, that we existed in a sea of light.

Now, since the light is already present everywhere in whatever volume, that is, since that light has already filled up the entire volume, there is nothing else that can similarly arise there that is not of the nature of light. Even if something else were to arise, being surrounded by light it would eventually succumb to that light. So the light is all-powerful or omnipotent within that volume.

Further, since the light exists simultaneously everywhere in that space and has a complete knowledge of itself it therefore has a complete knowledge of that space or of anything that can arise in that space. So it is all-knowing or omniscient in that space.

Finally, the light connects everything together instantaneously and holds these connections and the things connected in its nature, so it is all-nurturing or omninurturing within that space.

So, as a result of the infinite speed of light, light appears to have properties of omnipresence, omnipotence, omniscience, and omninurturing.

The Big Bang, The Seed Equation, & Entanglement

We hear about the Big Bang as the start of the universe. In this Big Bang matter is created. But in the examinations just presented the creation of matter is nothing other than the result of light traveling at the finite speed, and in this universe at 186,000 miles per second in vacuum. So we can say that the Big Bang, the apparent start of the universe, is the result of a slowing down of light from an infinite speed to some finite speed.

When light slows down then energy accumulates in packets or quanta and this results in what was inexpressible being able to express itself as matter. This can be thought of in terms of an incredibly rapidly moving stream of water. If the water is traveling so fast then no boundary will be able to contain it and the energy will be continuous over the length of the stream. If it is traveling slower though, then the water will be able to be held by boundaries along the length of the stream. The energy in this case will be discontinuous and will appear in "packets".

In this process of slowing down the implicit omnipresent-omnipotent-omniscient-omninurturing nature of light becomes or transforms into an explicit or emergent matter-past-present-future nature of light. So there is implicit in this transformation a high degree of 'entanglement' as it were. That is, ab initio, everything that appears in this universe is highly entangled. In other words every fundamental particle, whether it be a boson or lepton, or any subsequent emergence of matter is already deeply entangled by virtue of having emanated from the same single starting-point conceptualized as the Big Bang. This process of transformation creates a seed-equation that is present and synonymous with the Big Bang. The dynamics of a resulting universe are contained in the dynamics of such a seed-equation. The seed-equation will be explored in more detail in the section on mathematical foundations.

We have arisen in the field of light that has a finite speed. It is difficult therefore to feel the reality of the omnipresent-omnipotent-omniscient-omninurturing light. We are attuned to perceiving in the material world that has resulted due to the finite speed of light. But if we could step back into the fullness of light in its pure state at infinity from there we would likely see that there can be different universes created as a result of light selectively slowing down to different levels. The slowing down to a different level will create a particular kind of universe.

Note that there have been experiments to slow down light so that it practically moves at a snail's pace. This slowing down has to be put into context. Even if light were to slow down to 1 mile per second, say, from 186,000 miles per second, keep in mind that it is possible that this change in speed is likely only a miniscule fraction of the change in speed from light traveling infinitely fast to light traveling at c, that is, 186,000 miles per second. So in effect what could be said is that when light is made to slow down in experiments, the range or band that it slows down to deviates only slightly from its relative-to-infinity slower speed in vacuum.

Necessity of Constant Speed of Light and Variable Time and Space

The constant speed of light, at a speed less than an infinite speed, allows a build up of energy at quantum levels, that in turn allows any properties or possibility in light to express itself materially. When light travels at the speed of c – 186,000 miles per second – then the result of that is the material universe as we see and experience it, also with its division of time into a past, a present, and a future. In fact, we could say that for the known universe, as we experience it, light had to travel at c for it to arise. The speed of light had to be constant else there would be a variable, fluid reality to matter, likely displaying barely forming islands of matter subject to sudden disappearance, reappearance, continuing ad infinitum.

But also since the distance traveled by an object or a ray is the result of the speed it is traveling at for the time involved, this gives us an interesting insight that Einstein based his Theory of Relativity on (Einstein, 1995). Since the speed of light is constant, and has to be for the universe to be in its observable stable condition, this means time and distance (or space) potentially can vary. So if an object is traveling very fast, then time and space are going to be experienced differently by it, as compared with an object that is traveling considerably slower: as the speed of an object approaches c time slows down and distance contracts.

So an object that manages to travel at a speed of c will in some sense partake of the reality as experienced by c. It will transcend the conventions of time and space that are set up because of c and experience these differently.

Light at Speed Close to Zero

On the flip-side imagine light traveling at a speed close to or approaching zero. In this case the experienced reality is going to be the opposite of reality as experienced when light is traveling at an infinite speed.

First of all, light emanating from any point will remain fundamentally isolated and become the basis of extreme fragmentation. This has to be since regardless of the amount of elapsed time the light will still be only at its point of origin or source. So instead of a Presence of Light in any considered volume, there will be an Absence of Light.

Further, in imagining an area or volume, such light will have no knowledge of what is going on in the volume, and hence be completely ignorant, will further, have no power to effect any circumstances and hence be completely powerless, and finally, will not be able to coordinate or be present in any relationships. Hence it will be detrimental to or agnostic of any relationships instead of harmonious or nurturing to relationships.

Hence, reality in such a scenario would be stark, dark, desolate, fruitless, completely fragmented, and even perhaps pointless.

Such a reality may represent a negative-infinity and may be thought of as an extreme limit of what may be possible in worlds in which Light projects itself.

Seen from this point of view a Big Bang would be a fortuitous event since it will allow a fruitful reintegration of different worlds, each created by light traveling or projected at a different speed, to perhaps begin to take place as will be discussed further.

Understanding Superposition In A Light-Based Model

Light traveling at different speeds can, as illustrated by the cases of an infinite speed, c, and 0, create different realities. But if the infinite speed case is the native state, and if every other case is a projected state, then it is possible that these realities all exist simultaneously. In other words, phenomena resident in each reality created by the native state and by multiple projected states can be superposed.

Further, as will be explored in detail in subsequent chapters such states of superposition must be considered as superposition because they are inherently related to each other. A phenomenon that occurs in the reality where light travels at c, has its origin in phenomena or functionality existing in reality where light travels at a speed greater than c, and is influenced by phenomena existing in a reality where light travels at 0 miles per second.

This notion of superposition is an alternative to the currently conceived notion of superposition common to the Copenhagen Interpretation of Quantum Mechanics where a quantum object can simultaneously exist in infinite possible states until it is measured. Such a light-based interpretation of superposition also implies a fundamentally different model of computation at the quantum level, and therefore a completely different view of the possibilities of quantum computing.

Chapter 1.2: Light & Quanta

This chapter explores quanta, its relation to light in more detail, and other emergent properties that must also be quantized because of multiple speeds of light.

What is Light & What are Quanta?

When we think of light traveling at the speed of c, there are properties in the reality that emerge as a result of this. These properties as we have examined are of the nature of matter, the past, the present, and the future. These properties are the result of light traveling at c and so can be thought of as emerging from light.

But what do these properties actually mean? And further, how do these properties relate to the apparently quite different properties of omnipresence, omnipotent, omniscience, omninurturing that can appear in a reality where the speed of light is infinite?

Let us consider first the properties of light that emerge when light is traveling at c: the past, the present, the future, and matter.

What is the past? It is the perceivable result of all the work and effort that has taken place so far. It is the foundation upon which the present and future will be built. It represents a status quo, a stability, and even a rigidity, and given that it is the result of the long play of time, it will not easily be persuaded to become another thing. It can be thought of as that which the eye can see when it looks around it. There is "physicality" to what the eye can see and so the essence of the past is a kind of physical-ness. So, ingrained in light, is this ability to project or create physical-ness.

What is the present? It is the tremendous play of forces of all kinds to express themselves here and now. There is "vitality" that is present in this play and often it is the most energetic or forceful of the forces that will win out, as opposed to the most insightful or thoughtful. All the tremendous possibility of the future is seeking for expression now and so this essential vitality can also be thought of as a projection or possibility implicit in light.

What is the future? It is the inevitability of what will manifest. The great thoughts, the great ideas, the purpose, the possibilities will sooner or later express themselves in what we call the future. And the essence of this is thoughtfulness or a curiosity or a purpose that we can summarize as an essential "mentality". So embedded in light is this ability to project mentality.

And what is matter? It is the myriad crystallization into apparent diversity of the one essential reality of light, to allow for a play between these different sides or possibilities in an increasingly harmonious interaction. So its essence is "harmony", and this too can be thought of as an essential property projected from or implicit in light.

But what about when light is considered to move at an infinite speed? Then the properties that become apparent, as already explored, are those of omnipresence, omnipotence, omniscience, and omninurturing. But omnipresence has physical-ness to it, omnipotence has vitality to it, omniscience a mentality to it, and omninurturing a harmony to it. It may be said that physical-ness comes from omnipresence, vitality comes from omnipotence, mentality comes from omniscience, and matter comes from omninurturing.

So whether light is traveling at an infinite speed or the speed we know as c, there is something about the properties it projects, that in essence is the same. So let us refer to these essential properties made apparent through the worlds that are created, as Presence (from omnipresence), Power (from omnipotence), Knowledge (from omniscience), and Harmony (from omninurturing).

Further, quanta can be thought of as existing on that very border of the world or reality created due to Light traveling at c, and realities created as Light travels faster, which at its limit is an infinite speed. So quanta are a doorway or an interface into worlds of Light, and a doorway by which possibilities in deeper worlds of Light can express themselves here in this material realm. Being so, phenomenon such as superposition and entanglement are natural to quanta. But superposition as has just been preliminarily explored, is a supra-physical phenomenon. Entanglement while apparent in the Big Bang, and therefore a physical phenomenon, can also occur in anterior light-realities, thereby also becoming a supra-physical phenomenon in addition to a physical phenomenon.

The model of Light explored in subsequent chapters will mathematically explore these notions of supra-physical superposition and supra-physical entanglement in more detail.

Structured Time, Structured Space, and Their Relation To Quanta

In considering the worlds that are created due to the way light moves or the play of light, we see properties that are projected because of it. In the finite world, the world that results from light traveling at the speed c, there is, relatively, smallness we can grasp and on which we can build. In the infinite world that results from light traveling at an infinite speed there is a vastness and fullness that is difficult to grasp.

The notions of time and space are something entirely different in both worlds, and it could be said that Space allows the full play of everything meant by Power, Knowledge, Harmony, and Presence to be seeded in it, and that Time allows that seeding to flower into fuller forms with its passage.

In other words both space and time are not just abstract concepts but are essentially highly structured to allow physical-ness, vitality, mentality, and matter to become Presence, Power, Knowledge, and Harmony.

Space allows all the possibilities present in Presence, Power, Knowledge, and Harmony and seeded in vast diversity, to evolve into more fullness through the time stages of physical-ness, vitality, mentality, and evolving matter.

Such a creation of space and time is synonymous with the creation of quanta. Quanta become the means for the possibilities inherent in the anterior worlds of Presence, Power, Knowledge, and Harmony to express themselves in a structured space and time. Quanta are therefore a passage into deeper worlds of Light, and a means for possibilities in these deeper realms to express themselves materially.

But further, in this view it may also be proposed that space and time being so structured by the four properties of Light, need also to be quantized. That is, space and time must be experienced as quanta as well.

The Quantization of Space, Time, Matter, and Gravity

Chapter 1.1 proposed that quanta is the result of the slowing down of light, and as such becomes the basis for material expression. In other words, for matter to express itself requires quanta. But further it was also just proposed that space, consisting of seeds of the properties of light, would also require quanta to express itself. Philosophically this has to be in a world where light is traveling at a fraction of its possible speed. Further, it was also proposed that time is highly structured, expressing the growth of the seeds in space through definite phases. As such, growth through such phases can also be thought of as happening due to quantization that so allows phases to express themselves.

But if we take a deeper look at space, time, and matter in light of the properties of Light, it can be deduced that space, consisting of a vast array of seeds derived from the properties of Light is itself an expression of Light's property of Knowledge.

Time, bringing forth the meaning contained in the seeds, regardless of circumstance, and even being opposed by circumstance, can be thought of as Light's property of Power.

Matter itself, being a container in which space and time can allow deeper properties of Light to become materially tangible, must be an expression of Light's property of Presence.

But it is also known from Einstein's General Theory of Relativity that gravity is associated with mass and space, in that, as the physicist John Wheeler has put it, it is none other than a mass's instruction telling space how to curve, and again is nothing else that space's instruction telling mass how to move through it (Wheeler, 2000). As such, where mass and space exist, there gravity has to exist as well. Hence it may be inferred that gravity is none other than an expression of Light's property of Harmony, which fixes the collective relationship between object and object.

But if as proposed, matter, space, and time all need to be quantized in order to express themselves, then this must also be true of gravity.

But also, the emergence of space, time, matter, and gravity can be seen as a quantum computational outcome of light precipitating from the reality where it travels at an infinite speed to the reality where it travels at c.

Subsequent chapters will explore these claims in greater detail.

This chapter will explore some issues that lay doubt on the current foundations of quantum computing.

Superposition

The notion of superposition that is the foundation of modern day quantum computing has ironically never been adequately proven. It is based on the Copenhagen Interpretation that theorizes that until a photon, or particle, or other quantum-level object is measured, it exists in an infinite number of superposed states that collapses into a perceivable state only through the act of measurement. But if that were the case, then there is no basis for the universe having come into being, or evolving the way it has, or producing a human species on planet earth, since there was conceivably never any measurement to have precipitated any of those outcomes in the first place.

If such outcomes could materialize, and the universe is the proof that it has, then the model on which the current notion of superposition is based, and therefore our very notion of superposition is likely incomplete. And this becomes important because further, the supposed extraordinary processing power of quantum computers is also based on this unproven or incomplete notion of superposition. By contrast classical computing proceeded along an extremely rigorous path with conceptualization and creation of bits and an entire mathematics and carefully designed computer languages to manipulate these bits to process information. The fact is that superposition may have an entirely different way of operating than has been conceived to date.

The Need for a Supra-Physical Process

Further the very stuff of the physical, including cosmic prerequisites of space, time, gravity, and energy, can be modeled as emergent from a reality of light. There is a supra-physical process that creates the physical. To therefore approach quantum computing in a purely 'physical' manner is ab initio likely incomplete. In fact classical computing itself has a supra-physical process, such as thought-based algorithms, applied to the physical layer of bits that has made it successful. For quantum computing to be successful, superposition, entanglement, and other essentially supra-physical phenomena usually classified as quantum 'weirdness', has similarly to become a repeatable process that needs to be consciously applied to, rather than 'collapsed' into the physical layer. The physical layer has to be such that it allows or records fullness of any subjectivity to be understood.

The Problem with Statistics and Probability

Further, commonly accepted quantum mechanics interpretations such as the Copenhagen and Ensemble Interpretations erect statistical and probability functions as co-equal with the nature of reality. That is, the nature of reality is taken to be statistical and probabilistic. This is justified by suggesting that while individual phenomenon are entirely probabilistic or random, in aggregate they become deterministic or predictable. But alternatively rather than such statistical aggregation, in the Light-centered interpretation of Quantum Mechanics suggested in this book, reality can be modeled as being "functional" as will be explained in greater detail, which then allows a process of *statistical disaggregation* to take place whereby apparent randomness can be seen as a bi-product of function-focused native light precipitating into more material layers that come about when light travels at c, as suggested in Chapter 9.3. Briefly, in such a functional-view of the universe the granularity of analyses cannot just be quantum-level particles as is currently the case, but quantum-functionality in which relationships, function, and therefore laws are implicit in the supposedly smallest phenomenon.

Summary

The very basis of modern-day quantum computing that relies on infinite number of superposed quantum states, on probability, on observable measurement that brings things into reality, is bought into question in the Light-centered Interpretation discussed in this book. In fact from the point of view of the latter interpretation superposition, entanglement, and reality take on a different meaning and the infinite processing power allegedly true of quantum states, simply does not exist in the manner in which it has been conceived.

SECTION 2: THE MATHEMATICAL FOUNDATION FOR A LIGHT-BASED INTERPRETATION OF QUANTUM PHENOMENA

This section explores the mathematics required to interpret the universe as a play of functional-richness. Note that an earlier version of the mathematics worked out in the author's doctoral thesis (Malik, 2017a) has been adapted and further developed in this book.

The Light-Matrix, explored in Chapter 2.1, describes the overarching structure from a ubiquitous point-instant to the vast diversity of life. The Light-Matrix also elaborates an alternative model of superposition to that held true from the Copenhagen Interpretation of Quantum Mechanics.

Each layer in the matrix is then explored in more detail in subsequent chapters through the point-instant, the architectural forces, uniqueness of organizations and its emergence, and the always-available summary inherent dynamics. Exploration of the dynamics of each layer also sheds insight into the phenomenon and variability of entanglement.

Further, dynamics of possibility are explored through qualified determinism. Qualified determinism presents the mathematics of selecting a likely outcome from the multi-layered superposition-cum-entanglement dynamics existing at the quantum level.

This section will hence, specifically explore the following foundational mathematics:
- Light-Matrix & Superposition
- Nature of Entanglement at a Point-Instant in a System
- Entanglement at the Level of Architectural Forces
- Uniqueness of Organizations
- Emergence of Uniqueness
- Inherent Dynamics of Any System
- Qualified Determinism

This chapter will explore the Light-Matrix, a mathematical expression of the codification in Light formed at the time of the Big Bang (Malik et al., 2018). This Light-Matrix can be thought of as a seed-equation that provides insight into dynamics of the universe. The Light-Matrix codifies dynamics that may be experienced at the quantum border of realities created through light traveling at different speeds, and further in the worlds that the quantum veil or window provide access to. Weaving together possible realities created through light traveling at different speeds, the Light-Matrix also provides a model of superposition.

Light slowing down or projecting itself at a particular speed is envisioned to create a particular type of universe that will exist in all universes where light is traveling at any faster projected speed. Hence, all universes will exist in that created where light is in its native state traveling at an infinite speed. This implies that all properties true of the native state will also be present in any sub-universe. Similarly, the universe created when light is projected at c, will have all properties created in universes created through light traveling faster than c, potentially acting in it. It is the presence of such properties and dynamics true of supra-universes that that gives rise to the phenomenon of superposition in any sub-universe.

Nature of Light at U

As laid out in Section 1, it is perhaps fair to say that the speed of light has significant implications on the experienced nature of reality. The finiteness, c, at 186,000 miles per second in a vacuum creates an upper bound to the speed with which any object may travel also implying, and as discussed in Section 1, that objective reality will be experienced as a past, a present, a future from the point of view of that object. These characteristics – a past, a present, a future – are implicit in the nature of light and become part of objective reality because of the speed of light. So the way we experience time seems to be determined by c.

Further, c also creates a lower bound when inverted ($1/c$) being proportional, or arguably even determining Planck's constant, h, that pegs the minimum amount of energy or quanta required for expression at the sub-atomic level. Planck's constant, h, pegging the amount of energy required for expression, therefore may allow matter to form as suggested by the physicist Lorentz a century ago (Lorentz, 1925). Note that Einstein postulated quanta as a fundamental property of light itself, rather than as something that arose in the interaction of light with matter as suggested by Max Planck (Isaacson, 2008). Hence, c contributes to or even establishes a reality of nature with a past, present, and future, to also be experienced as a phenomena of connection between seemingly independent islands of matter. This characteristic of 'connection' is therefore also proposed to be implicit in the nature of light and becomes part of objective reality because of the speed of light.

But as explored in Chapter 1.2, a 'past' can also be viewed as established reality as defined by what the eye or other lenses of perception can see. Hence, 'It is the perceivable result of all the work and effort that has taken place so far. It is the foundation upon which the present and future will be built. It represents a status quo, stability, and even rigidity, and given that it is the result of the long play of time, it will not easily be persuaded to become another thing. It can be thought of as that which the eye can see when it looks around it. There is "physicality" to what the eye can see and so the essence of the past is a kind of physical-ness. So, ingrained in light, is this ability to project or create physical-ness.'

Also as explored previously, the 'present' is 'the tremendous play of forces of all kinds to express themselves here and now. There is "vitality" that is present in this play and often it is the most energetic or forceful of the forces that will win out, as opposed to the most insightful or thoughtful. All the tremendous possibility of the future is seeking for expression now and so this essential vitality can also be thought of as a projection or possibility implicit in light.'

The 'future' is 'the inevitability of what will manifest. The great thoughts, the great ideas, the purpose, the possibilities will sooner or later express themselves in what we call the future. And the essence of this is thoughtfulness or a curiosity or a purpose that we can summarize as an essential "mentality". So embedded in light is this ability to project mentality.'

These implicit characteristics of the nature of light as experienced at the layer of reality so set up by a finite speed of light may hence be summarized by Equation 2.1.1, where c_U refers to the speed of light of 186,000 miles per second, that has created the perceived nature of reality, U, as expressed in Equation 2.1.1:

$$c_U : [Physical, Vital, Mental, Connection]$$

Eq. 2.1.1: Implicit Characteristics of the Nature of Light at U

Exploring further, it is thought however that at quantum levels the nature of reality is characterized by wave-particle duality. Light itself and matter may be experienced as both particles and waves. But for matter to be experienced as waves implies that 'h' must have become a fraction of itself, $h_{fraction}$, to allow the concentration or possibility of quanta to have dispersed into wave-form. This further implies that c must have become greater than itself, c_N, such that the inequality specified by Equation 2.1.2 holds:

$$c_N > c_U$$

Eq. 2.1.2: Layer N, Layer U Inequality of Speed of Light

Note that what is implied here is that just as there is a nature of reality specified by U that is the result of the speed of light being 186,000 miles per second, so too there is another nature of reality specified by N that is the result of a speed of light greater than 186,000 miles per second.

This is akin to recent developments in physics with the notion of property spaces being separate from but influencing physical space as explored by Nobel Physicist Frank Wilczek in his book 'A Beautiful Question" (Wilczek, 2016). But further in "Slow Light" Perkowitz's recent treatment of today's breakthroughs in the science of light (Perkowitz, 2011) he states: "Although relativity implies that it's impossible to accelerate an object to the speed of light, the theory may not disallow particles already moving at speed c or greater."

So light traveling at c_N may be possible. Current instrumentation, experience, and normal modes of thinking though, having developed as a bi-product of the characteristics so created in the layer of reality U may be inadequate to access N without appropriate modification. The notion of wave-particle duality already challenges the notion of normal thinking perhaps because wave-like phenomena could be a function of faster than c motion, and particle-like phenomena a function of equal to c motion. That these may be happening simultaneously is reinforced by principles such as complementarity in which experimental observation may allow measurement of one or another but not of both as pointed out by Whitaker (Whitaker, 2006). But further the very notion of the Pilot-Wave Interpretation of Quantum Mechanics, as modeled by the physicists Bohm and DeBroglie (Holland, 1995), is that constructive interference of waves causes particles to move toward these areas. Hence in this interpretation both particle and wave always exist simultaneously.

But then taking this trend of a possible increase in the speed of light to its limit, this will result in a speed of light of infinite miles per second. The question is, what is the nature of reality when light is traveling at infinite miles per second? As explored in Section 1, in any continuum light originating at any point will instantaneously have arrived at every other point. Hence light will have a full and immediate *presence* in that continuum. Further, that light will *know* everything that is happening in that continuum completely and instantaneously – that is know what is emerging, what is changing, what is diminishing, what may be connected to what, and so on - or have a quality of knowledge. It will connect every object in that continuum completely and therefore have a quality of connection or *harmony*. Finally nothing will be able to resist it or set up a separate reality that excludes it and hence it will have a quality of *power*.

These implicit characteristics of the nature of light as experienced at the layer of reality so set up by an infinite speed of light may hence be summarized by Equation 2.1.3, where c_∞ refers to the speed of light of ∞ miles per second, that has created the perceived nature of reality, ∞:

c_∞:[*Presence, Power, Knowledge, Harmony*]

Eq. 2.1.3: Implicit Characteristics of Nature of Light at ∞ Speed

Transformation from ∞ to U

But by (2.1.3) it can also be noticed that 'physical' is related to Presence, 'vital' is related to Power, 'mental' is related to Knowledge, and 'connection' is related to Harmony.

The question then, is how do these apparent qualities at ∞ precipitate or become the physical-vital-mental-connection based diversity experienced at U? This may be achieved through the intervention or action of a couple of mathematical transformations acting on the implicit characteristics of nature of light at ∞ speed as summarized by (2.1.3).

First, the essential characteristics of Presence, Power, Knowledge, Harmony that it is posited exists at every point-instant by virtue of the ubiquity of light at ∞ will need to be expressed as sets with up to infinite elements. Second, elements

35

in these sets will need to combine together in potentially infinite ways to create a myriad of seeds or signatures that then become the source of the immense diversity experienced at U. This suggests that all that is seen and experienced at U may be nothing other than 'information' or 'content' of light and as such that there are fundamental mathematical symmetries at play where everything at U is essentially the same thing that exists at ∞. The heart of creation, in this view, is likely a constant quantum-computation that interrelates various realities so set up by light traveling at different speeds.

Assuming that the first transformation occurs at a layer of reality K where the speed of light is c_K, such that $c_U < c_K < c_\infty$, this may be expressed by Equation 2.1.4:

$$c_K : [S_{Pr}, S_{Po}, S_K, S_H]$$

Eq. 2.1.4: The First Transformation at Layer K

S_{Pr} signifies 'Set of Presence', S_{Po} signifies 'Set of Power', S_K signifies 'Set of Knowledge', S_H signifies 'Set of Harmony/Nurturing' (note: Harmony and Nurturing will be used interchangeably through this book).

Assuming that the second transformation occurs at a layer of reality N where the speed of light is c_N, such that $c_U < c_N < c_K < c_\infty$, this may be expressed by Equation 2.1.5:

$$c_N : f(S_{Pr} \times S_{Po} \times S_K \times S_H)$$

Eq. 2.1.5: The Second Transformation at Layer N

The unique seeds are therefore a function, f, of some unique combination of the elements in the four sets S_{Pr}, S_{Po}, S_K, S_H.

The relationship between the layers of light may be modeled by the following matrix in Equation 2.1.6:

$$Light_{Matrix} = \begin{vmatrix} c_\infty:[Pr, Po, K, H] \\ (\downarrow R_{C_K} = f(R_{C_\infty})) \\ c_K: [S_{Pr}, S_{Po}, S_K, S_H] \\ (\downarrow R_{C_N} = f(R_{C_K})) \\ c_N: f(S_{Pr} \times S_{Po} \times S_K \times S_H) \\ (\downarrow R_{C_U} = f(R_{C_N})) \\ c_U: [P, V, M, C] \end{vmatrix}$$

Eq. 2.1.6: Light-Matrix

The matrix should be read from the top row down to the bottom row as indicated by the $\downarrow$ between rows, and suggests a series of transformations leading from the ubiquitous nature of light implicit in a point – presence, power, knowledge, harmony - to the seeming diversity of matter observed at the layer of reality U which is fundamentally the same presence, power, knowledge, and harmony projected into another form of itself.

The first transformation is summarized by Equation 2.1.7:

$$R_{C_K} = f(R_{C_\infty})$$

Eq. 2.1.7: Light-Matrix First Transformation

This is suggesting that the reality at the layer specified by the speed of light ^{C}K, R_{C_K}, is a function of the reality at the layer specified by the speed of light $^{C}\infty$. This transformation translates the essential nature of a point into the sets described in Equation 2.1.4.

The second transformation is summarized by Equation 2.1.8:

$$R_{C_N} = f(R_{C_K})$$

Eq. 2.1.8: Light-Matrix Second Transformation

This is suggesting that the reality at the layer specified by the speed of light ^{C}N, R_{C_N}, is a function of the reality at the layer specified by the speed of light ^{C}K. This transformation combines elements of the sets into unique seeds as suggested by Equation 2.1.5.

The third transformation is summarized by Equation 2.1.9:

$$R_{C_U} = f\left(R_{C_N}\right)$$

Eq. 2.1.9: Light-Matrix Third Transformation

This is suggesting that the reality at the layer specified by the speed of light ^{C_U}, R_{C_U}, is a function of the reality at the layer specified by the speed of light ^{C_N}. This transformation builds on the unique seeds suggested by (2.1.5) to create the diversity of U as specified by (2.1.1).

In this framework the notion of wave-particle duality hence may become complementary block-field-wave-particle "quadrality" where block refers to phenomenon resident to ∞, field to phenomenon resident to N, wave to phenomenon resident to K, and particle to phenomenon resident to U. The block is all the reality always present behind the surface and is captured by (2.1.3) or the top line in the Light-Matrix as expressed in (2.1.6). The field is captured by (2.1.4) or the second line from the top in (2.1.6) and can be thought of as layers of possibility existing in each of the sets. The wave is captured by the creation of seeds represented by (2.1.5) or by the third line in (2.1.6). The particle that is apparently disconnected from the whole is captured by (2.1.1) or the bottom line in (2.1.6).

So what we arrive at is a fundamental Light-Matrix that suggests key dynamics for different layers of Light. Quanta, being at the interface of at least Layer U and Layer N are so positioned as being subject to multiple sets of dynamics.

Layer Zero (0)

For the sake of completeness the thought-experiment to do with light traveling at a speed of zero must now be brought up. In Chapter 1.1 a possible reality that would result if light were to travel at this speed was described as opposite to the reality were light to travel at an infinite speed.

Hence, ubiquitous Presence of Light would become the ubiquitous Absence of Light or Darkness. The aspect of Power would become utter Weakness. The aspect of Knowledge would become complete Ignorance. The aspect of Harmony would become total Chaos.

And all this because Light would be unable to travel form where it is, in complete opposition to its known nature of traveling at a speed of c, and perhaps even at a speed of ∞. Light, hence, would be hidden in itself, so as to speak, in some sort of a negative infinity. An Equation, 2.1.10, Nature of Light at Speed Zero (0), would hence be:

$$c_0 : [Darkness, Weakness, Ignorance, Chaos]$$

Eq. 2.1.10: Nature of Light at Speed Zero (0)

Further, (2.1.6), the Light-Matrix, would be modified to become Equation 2.1.11, Light-Matrix with Zero-Limit, where c_0 implies light at zero-speed, and D, W, I, and C imply Darkness, Weakness, Ignorance, and Chaos respectively:

$$Light_{Matrix}(0_{limit}) = \begin{vmatrix} c_\infty : [Pr, Po, K, H] \\ (\downarrow R_{C_K} = f(R_{C_\infty})) \\ c_K : [S_{Pr}, S_{Po}, S_K, S_H] \\ (\downarrow R_{C_N} = f(R_{C_K})) \\ c_N : f(S_{Pr} \times S_{Po} \times S_K \times S_H) \\ (\downarrow R_{C_U} = f(R_{C_N})) \\ c_U : [P, V, M, C] \\ \Uparrow \\ c_0 : [D, W, I, C] \end{vmatrix}$$

Eq. 2.1.11: Light-Matrix with Zero-Limit

The implication of the bottom zero-limit line is that just as there is proposed to be an influence from the upper layers of light on U, as will be explored in subsequent chapters, so too there is a subtle influence from this lower layer of "light" that perhaps creates a certain obstinacy of the untransformed (U) nature of practical reality. Such obstinacy will be further explored in Chapters 2.5, 2.6, and 2.7 and perhaps reinforces the need for the notion of 'transformation' in the first place.

In subsequent chapters the Light-Matrix will assume the form as depicted in (2.1.6) as opposed to (2.1.11). The influence of the zero-limit will however be supposed in the operation of Layer U, which therefore, as just proposed will require the necessity of a continuous transformation in order to allow increasing functional-richness to manifest. Such an assumption is consistent with observation and as will be explored therefore requires human faculties such as will, thought, love, amongst others in order to counter the influence of darkness, ignorance, weakness, and chaos.

Superposition, Entanglement, and Quantum Computation in the Light-Matrix

As may be apparent in the Light-Matrix (2.1.6) and the complete form (2.1.11) circumstance at U is the outcome of a number of influences from layers ∞, K, N, U itself, and Zero (0). But further since each of these layers is itself a reality caused by a different speed of light, the interface between these layers occurs at the quantum-level. Hence the quantum-level is replete with superposition emanating from **layers** ∞, K, N, U itself, and Zero (0). There is simultaneity and multiple possibilities that will determine what manifests at U. But as will be explored in more detail in subsequent chapters there is a logic or process to what manifests.

The notion therefore that a quantum-object – an entity at the quantum-level – can be in an infinite number of superposed states as is presumed in several leading interpretations of quantum mechanics and subsequently in the current popular notion of quantum computing, is called into question. There may in fact be infinite number of possibilities vying for manifestation, but there may be a more precise process by which this manifestation takes place.

Further, entanglement exists ab initio at the layer ∞. This entanglement will be referred to as ∞-entanglement and will be further explored in Chapter 2.2. And since this layer in a sense contains every other layer entanglement is a reality of every quantum-object. Further, there are additional forms of entanglement that occur even before a quantum-object becomes perceivable at layer U. There is entanglement at layer K by virtue of field-type action of architectural sets. This will be referred to as K-entanglement and will be further explored in Chapter 2.3. There is an entanglement at layer N by virtue of the wave-type action of seeds. This will be referred to as N-entanglement and will be further explored in Chapter 2.4.

Quantum computation is therefore a reality in the manifestation of the minutest of circumstances at U. Everything is the result of a quantum-computation. Implicit in these computations are the dynamics of superposition, entanglement, and precipitation of functionality from one layer to another as will be explored in greater detail.

Chapter 2.2: ∞-Entanglement - Nature of a Point-Instant in a System

The 'point-instant' captures the inherent nature that appears to exist in the system and is represented by the top-line in Equation (2.1.6), the Light-Matrix. The point-instant is a function of the dynamics of that reality where light travels at ∞ speed and is envisioned as being infinitely entangled by virtue of light being omnipresent-omnipotent-omniscient-omninurturing in that realm. In other words this nature is a function of the properties of light derived in Equation (2.1,3) - Presence, Power, Knowledge, and Harmony – reproduced here for convenience:

$$c_\infty:[Presence, Power, Knowledge, Harmony]$$

The 'point' aspect of the point-instant suggests the space-dimension and that space is seeded with the possibilities inherent in the properties of Presence, Power, Knowledge, and Harmony. The 'instant' aspect suggests the time-dimension and gives insight into the process of emergence that the possibilities in space progressively surface as.

In its point-instant, Presence-Power-Knowledge-Harmony wholeness, Presence allows emergences to continue to develop as per the possibilities implicit in the past-present-future or physical-vital-mental pathway.

Beginning to translate this into an equation, the notation $System_{Pr}$ is given to system-presence. This system-presence is true across any considered Time-Space continuum starting from a time-space boundary '0' to a time-space boundary 'N'. This notion is characterized by the notation $TS_{0 \to N}$. Within that boundary from 0 to 'N', the 'presence' is such that it will always seize an opportunity to cause a shift from the physical-leading to the vital-leading, and from the vital-leading to the mental-leading. Research shows (Malik, 2009) that greater degrees of freedom is afforded by such a shift.

The notion that the 'presence' seizes on 'opportunity' is characterized by the notation:

$$\begin{array}{c} Presence \\ \downarrow \\ Opportunity \end{array}$$

The shift from physical-leading (P_L) to vital-leading (V_L) and vital-leading (V_L) to mental-leading (M_L) is characterized by:

$$P_L \;\rightarrow\; V_L$$
$$V_L \;\rightarrow\; M_L$$

Hence in this approach it is suggested that:

$$System_{Pr} \equiv TS_{0 \rightarrow N} \begin{bmatrix} Presence \\ \downarrow \\ Opportunity \end{bmatrix} \begin{bmatrix} P_L & \rightarrow & V_L \\ V_L & \rightarrow & M_L \end{bmatrix}$$

But there is something else about this Presence as well. All other developments take place in it. That is, it provides a container of sorts in which the plays of system-power, system-knowledge, and system-harmony/system-nurturing can take place. This notion is summarized by the notation:

$$Container \begin{bmatrix} System_P \\ System_K \\ System_N \end{bmatrix}$$

Hence, combining these various components, an equation for 'system-presence', Equation 2.2.1, arises:

$$System_{Pr} \equiv TS_{0 \rightarrow N} \begin{bmatrix} Presence \\ \downarrow \\ Opportunity \end{bmatrix} \begin{bmatrix} P_L & \rightarrow & V_L \\ V_L & \rightarrow & M_L \end{bmatrix} \& \; Container \begin{bmatrix} System_P \\ System_K \\ System_N \end{bmatrix}$$

Eq 2.2.1: System Presence

In its point-instant, Presence-Power-Knowledge-Harmony wholeness, Power allows emergences to continue to happen in spite of tremendous oppositions of all kinds; this too, regardless of field or area.

Constructing an equation for system-power, the notation $System_P$ is used to represent system-power. Any endeavor will always be met with resistances of various kinds. The resistances that arise along the physical dimension are referred to as P_R. The resistances that arise along the vital dimension are referred to as V_R. The resistances that arise along the mental dimension are referred to as M_R. In the fruition of any endeavor one or all of these types of resistances may arise. Further, resistance of one kind often feeds on resistance of another kind, and to generalize the resistances encountered in an endeavor, these may be characterized as the product of the three types of resistance:

$$P_R \; * \; V_R \; * \; M_R$$

These resistances arise across any considered Time-Space boundary from 0 to 'N', and therefore it may be said that the power of the system is such that:

$$power > \sum_{TS=0}^{N} P_R \; * \; V_R \; * \; M_R$$

An equation for 'system-power', Equation 2.2.2, hence, is the following:

$$System_P \equiv power > \sum_{TS=0}^{N} P_R \; * \; V_R \; * \; M_R$$

Eq 2.2.2: System Power

In its point-instant, Presence-Power-Knowledge-Harmony wholeness, Knowledge orchestrates emergences to continue to happen by leveraging the right instruments and circumstances.

Translating this into an equation, the notation, $System_K$, is used for system-knowledge. This $System_K$ is such that it leverages the right instrumentation and circumstance to bring about the progress that is possible. This concept of 'instrumentation' is denoted by the subscript 'I'. The concept of 'circumstance' is denoted by the subscript 'C'. Both instrumentation and circumstance can be of a physical, vital, or mental type and this possibility is denoted by:

$$\begin{bmatrix} P_{I,C} \\ V_{I,C} \\ M_{I,C} \end{bmatrix}$$

Further, the notion that the 'knowledge' is such that it 'leverages' the right instrumentation and circumstance is depicted by:

Knowledge
↓
Leverage

This act of leveraging results in a fundamental shift so that the physical-leading yields to the vital-leading, and the vital-leading yields to the mental-leading. Hence:

$$\begin{matrix} Knowledge \\ \downarrow \\ Leverage \end{matrix} \begin{bmatrix} P_{I,C} \\ V_{I,C} \\ M_{I,C} \end{bmatrix} \rightarrow \begin{bmatrix} P_L & \rightarrow & V_L \\ V_L & \rightarrow & M_L \end{bmatrix}$$

Since this behavior may exist across any Time-Space continuum an equation for system-knowledge, Equation 2.2.3, is suggested:

$$System_K \equiv TS_{0 \rightarrow N} \begin{bmatrix} Knowledge \\ \downarrow \\ Leverage \end{bmatrix} \begin{bmatrix} P_{I,C} \\ V_{I,C} \\ M_{I,C} \end{bmatrix} \rightarrow \begin{bmatrix} P_L & \rightarrow & V_L \\ V_L & \rightarrow & M_L \end{bmatrix}$$

Eq 2.2.3: System Knowledge

In its point-instant, Presence-Power-Knowledge-Harmony wholeness, Harmony or nurturing allows emergences to continue to happen with more and more degrees of freedom coming to the surface.

The characteristic of this implicit-nurturing may be referred to as 'system-nurturing'. Like the other characteristics it is suggested to exist across a Time-Space continuum. This is depicted by:

$$TS_{0 \rightarrow N}$$

There is an action of nurturing such that any state is always advanced to a higher level. This is depicted by:

$$\coprod_{Nurturing} \begin{pmatrix} P_- & M_+ \\ V_- & V_+ \\ M_- & P_+ \end{pmatrix}$$

Hence, there is a 'union', depicted by 'U' that 'nurtures' the negatives towards their positives.

Further, there is an increasing action of nurturing such that the possibility of integration is always increased to form a larger and larger basis. This increasing basis is depicted as being modulated by the polar coordinates 'r' and 'θ', where r is the radius which increases from an initial value of '0', and 'θ' is an angle from '0' to '360'.

Hence, the equation of system-nurturing, Equation 2.2.4, is depicted as:

$$System_N \equiv TS_{0 \to N} \left(\coprod_{Nurturing} \begin{pmatrix} P_- & M_+ \\ V_- & V_+ \\ M_- & P_+ \end{pmatrix} mod\ (r, \theta) \right)$$

Eq 2.2.4: System Nurturing

It is suggested that these four characteristics exist across any system, and to denote this it is generalized that every point in any system is embedded with this four-fold intelligence. It is suggested that this four-fold intelligence is resident in every instant-spot of the system. It is suggested that to be able to leverage or activate this four-fold intelligence at will is the ultimate act of innovation.

Note that while this state is presumed to be active in the system itself in every point-instant, and therefore the system as a whole has a logic that may be beyond normal comprehension, in subsequent chapters we will further explore the conditions for consciously activating the state of ∞-entanglement. A corollary of this already ever-active state is that entanglement is already a practical reality, and not just at the quantum levels.

Chapter 2.3: K-Entanglement - Architectural Forces

The characteristics embedded in a point as in (2.1.3) suggest a possibility that is hard to fathom. One can only glimpse the extraordinary nature embedded in a point. And yet it can be suggested that this extraordinary nature is barely visible unless the right analytical lens of the sort being suggested here is first set up. Further, it is suggested that this extraordinary nature is responsible for a broader set of architectural forces that exist behind the visible face of things.

Hence, system-presence, system-power, system-knowledge, and system-nurturing that define the nature of every point in our system, become more tangible as a broader set of architectural forces that emanate from each of them.

Considering system-presence, here is a characteristic that appears to be everywhere (Malik, 2009) at the service of all the constructs that develop within it. There is a diligence and perseverance by which any opportunity for progress is seized. Further, if one considers the extraordinary detail that appears in any construct, whether an atom, a body, a planet, or a galaxy, one is struck by the high degree of perfection that surfaces in this presence.

So if one contemplates the nature of this system-presence there is a set of forces that surface. Depicting such a set as $S_{System_{Pr}}$, one can arrive at elements such as Service, Perfection, Diligence, Perseverance, amongst others, that are part of this set. Hence, the set can be described by Equation 2.3.1:

$$S_{System_{Pr}} \ni [Service, Perfection, Diligence, Perseverance, ...]$$

Eq 2.3.1: Set of System Presence

But from the point of view of the speed of light, K is a reality set up by light traveling at c_K, where $c_U < c_N < c_K < c_\infty$. Conversely assuming an inverse proportionality with 'h', as c_K is closer to ∞, h will be closer to zero, and 'matter' or form will be highly dispersed. This dispersal is presumed to be field-like so that in effect any element of the set described in (2.3.1) or in the subsequent sets (2.3.2-4) will have a field-like reality and express k-entanglement such that all emergences emanating or comprised of that element will automatically partake in an entanglement with every other unique emergence having that element as part of its foundation. K-entanglement is therefore different from ∞-entanglement and every emergence will at least have both these types of entanglements subtly coordinating or influencing its action.

Similarly, considering the characteristic of system-power, one can hypothesize that there is a family of forces that emanates from it. The kinds of forces may be thought of as Power, Courage, Adventure, Justice, amongst others. The set for system-power can hence be depicted by Equation 2.3.2:

$$S_{System_p} \ni [Power, Courage, Adventure, Justice, ...]$$

Eq 2.3.2: Set of System Power

Similarly, considering the system-knowledge as the root of various powers that emanate from it, one may characterize the set for system-knowledge by Equation 2.3.3:

$$S_{System_K} \ni [Wisdom, Law Making, Spread of Knowledge...]$$

Eq 2.3.3: Set of System Knowledge

The set for system-nurturing is depicted by Equation 2.3.4:

$$S_{System_N} \ni [Love, Compassion, Harmony, Relationship ...]$$

Eq 2.3.4: Set of System Nurturing

Chapter 2.4 describes how unique signatures or seeds for any type of organization can be built by leveraging the elements in the sets of architectural forces.

Chapter 2.4: N-Entanglement - Uniqueness of Organizations

The hypothesis is that every organization, whether an atom, cell, person, team, corporation, market, or country is unique and that this uniqueness can be specified in terms of elements of the derived sets for power, knowledge, presence, and nurturing. This hypothesis for uniqueness stems from observations at multiple levels.

At the sub-atomic level Nobel Laureate Wolfgang Pauli's 'Pauli Exclusion Principle' states that no two similar fermions, which include fundamental particles with half-integer spin such as protons, neutrons, and electrons, can occupy the same quantum states simultaneously (Pauli, 1964). Spin has to do with the angle that the particle has to rotate through before being symmetrical with its original state. Half-integer spin particles need to rotate through 720 degrees before being symmetrical with their original state. The implication of the Pauli Exclusion Principle is that fundamental structure and consequently stability comes into being at the atomic level, which as is evident in the periodic table allows the separation of function related to form. This stability related to the underlying structure of atoms implies the basis of uniqueness and diversity. In the absence of the Exclusion Principle matter would just be a dense soup (Hawking, 1988) with particles occupying overlapping space.

At the observable level uniqueness is evident from the immense diversity of distinct species on earth (Mora, 2011) estimated to be over 2 million, and further the uniqueness of every member of each species. This member-level uniqueness is suggested by the difference in non-coding regions of the DNA that may vary in their sequence by about 1 to 4 percent, which in turn result in unique protein binding sequences of each human (Snyder, 2010), as an example, which in turn results in unique observable qualities.

At the astronomical level Einstein's Special Theory of Relativity (Einstein, 1995) suggests that every coordinate system potentially has its own space-time rendering as opposed to there being one absolute space and time. This implies the notion of uniqueness as an implicit property of space.

The four properties explored in the Chapter 2.1 and 2.2 define the source of that uniqueness. From this source emanate 4 sets of forces that suggest the boundaries of that uniqueness.

Assuming then that the fount of uniqueness is system-presence, a general equation for organizations that belong to the family of system-presence can be derived. Such uniqueness can be depicted as Sig_x where the subscript 'x' refers to the source family, and 'Sig' or signature to 'uniqueness'. Hence the uniqueness of an organization in the family of system-presence would be notated by $Sig_{System_{Pr}}$.

In line with the development of properties of a point and the precipitating architectural forces as discussed in Chapter 2.1 and 2.2 respectively, an approach to constructing such uniqueness is to assume a primary factor X that drives the uniqueness that belongs to the set $S_{System_{Pr}}$. Further, assume that the uniqueness is qualified by a number of secondary factors Y that may belong to any of the 4 sets - $S_{System_{Pr}}, S_{System_P}, S_{System_K}, S_{System_N}$. The primary factor X would have a greater weightage than any of the secondary factors Y. The weightage of X hence could be depicted by the number 'a', and the weightage of Y a number $'b_{0-n}'$, such that a > b. Further, the secondary element can repeat from '0 – n' times, and is hence depicted as $Y\bar{b}_{0-n}$.

The equation, Equation 2.4.1, hence for a unique organization derived from the family of system-presence is:

$$Sig_{Pr} = Xa + Y\bar{b}_{0-n} \ where \begin{bmatrix} X \in [S_{System_{Pr}}] \\ Y \in [S_{System_{Pr}}, S_{System_P}, S_{System_K}, S_{System_N}] \\ a, b \ are \ integers; a > b \end{bmatrix}$$

Eq 2.4.1: System Presence Based Unique Organization

But from the point of view of the speed of light, N is a reality set up by light traveling at c_N, where $c_U < c_N < c_K < c_\infty$. Conversely assuming an inverse proportionality with 'h', as c_N is closer to but greater than c_U, h_U will be less than h, and 'matter' or form will be unable to accumulate as it does at U, instead tending to disperse like a wave. The wave-like nature implies an entanglement, N-entanglement, such that any emergence will always be unique. Uniqueness could not be unless there was a dynamic such as N-entanglement in place.

Similarly, an equation, Equation 2.4.2, for a unique organization derived from the family of system-power is:

$$Sig_P = Xa + Y\bar{b}_{0-n} \quad where \begin{bmatrix} X \in [S_{System_P}] \\ Y \in [S_{System_{Pr}}, S_{System_P}, S_{System_K}, S_{System_N}] \\ a, b \ are \ integers; a > b \end{bmatrix}$$

Eq 2.4.2: System Power Based Unique Organization

An equation, Equation 2.4.3, for a unique organization derived from the family of system-knowledge is:

$$Sig_K = Xa + Y\bar{b}_{0-n} \quad where \begin{bmatrix} X \in [S_{System_K}] \\ Y \in [S_{System_{Pr}}, S_{System_P}, S_{System_K}, S_{System_N}] \\ a, b \ are \ integers; a > b \end{bmatrix}$$

Eq 2.4.3: System Knowledge Based Unique Organization

An equation, Equation 2.4.4, for a unique organization derived from the family of system-nurturing is:

$$Sig_N = Xa + Y\bar{b}_{0-n} \quad where \begin{bmatrix} X \in [S_{System_N}] \\ Y \in [S_{System_{Pr}}, S_{System_P}, S_{System_K}, S_{System_N}] \\ a, b \ are \ integers; a > b \end{bmatrix}$$

Eq 2.4.4: System Nurturing Based Unique Organization

The four preceding equations can be generalized by Equation 2.4.5:

$$Sig = Xa + Y\bar{b}_{0-n} \quad where \begin{bmatrix} X \in [S_{System_{Pr}}, S_{System_P}, S_{System_K}, S_{System_N}] \\ Y \in [S_{System_{Pr}}, S_{System_P}, S_{System_K}, S_{System_N}] \\ a, b \ are \ integers; a > b \end{bmatrix}$$

Eq 2.4.5: Generalized Equation for Unique Organization

Having considered the structure of uniqueness, the next question is how does such uniqueness emerge? This is discussed in Chapter 2.5.

Chapter 2.5: Emergence of Uniqueness

While the uniqueness of organizations as represented by the Signature is a seed, like any seed there is a process for its emergence (Kaufmann, 1995; Portugali, 2012; Yates, 2012), and the uniqueness will often be hidden or very much behind the scene until certain conditions are fulfilled (Malik, 2009).

The implicit nature of Time and Space suggest a universal developmental model that provides a cue as to the process for emergence (Deep Order Mathematics Videos, 2016). In this model the four sets of architectural forces already described form a pool in space, as it were, from which possibility arises. Possibility itself is unique from point to point and is governed by the Equation for Uniqueness (Equations 2.4.1 through 2.4.5) described in the previous chapter.

In other words, Space contains seeds, as suggested in Section 1, and as just discussed in the previous chapters in Section 2, seeds can be thought of as the result of the superposition of ever-present ∞-entanglement, K-entanglement, and N-entanglement, Thus it could be said that Space is filled with multiple levels of superposed entanglements in static form. Time on the other hand appears to be the working out of the possibilities implicit in these superposed entanglements. In equation form Space could be described by Equation 2.5.1:

$Space =$

$STATIC(superposition\ (\infty - entanglement, K - entanglement, N - entanglement)$

Eq 2.5.1: Space

In equation form Time could be described by Equation 2.5.2:

$Time =$

$DYNAMIC(superposition\ (\infty - entanglement, K - entanglement, N - entanglement)$

Eq 2.5.2: Time

Hence it is observed that initially the uniqueness takes a 'physical' form, moving on to a 'vital' form, and then onto a 'mental' form. Once the orientations implicit in each of these phases are assimilated, then the uniqueness takes on an 'integral' form. The integral form is a threshold phase, and allows the uniqueness suggested by the Signature to emerge in fuller force or in its 'force' form. The final phase is the 'contextual form' that allows the signature to act with impunity within a considered context.

Mathematically, if an organization exists at the physical phase, it may be suggested that its signature or uniqueness is modulated by the constant $'\pi'$. π is the seed of a circle or sphere and can be thought of as defining behavior that is tightly bound. Within such a tightly bound volume it will likely not even be apparent what the uniqueness of an organization necessarily is. Assuming the uniqueness to be defined by the derived equation *Sig*, the physical-level (P) behavior can be described by the following equation-segment where 'mod' signifies modulated-by:

P: $Sig * mod\ (\pi)$

If an organization exists at the vital level, it may be suggested that its uniqueness is modulated by the Euler-constant 'e'. e is at the root of exponential behavior. The vital by definition is about assertive and aggressive growth the symbol of which is 'e'. Hence vital-level (V) modulation (represented by 'mod') can be described by the following equation-segment:

V: $Sig * mod\ (e)$

If an organization exists at the mental level, it may be suggested that its uniqueness is modulated by the Gaussian Distribution 'G'. G summarizes rational behavior with a key direction followed by most, and directions more on the edge followed by outliers. Mental-level dynamics are arguably quite similar, and it can be suggested are best modeled by such a distribution (Salkind, 2007). Mental-level (M) modulation (mod) can hence be described by the following equation-segment:

M: $Sig * mod\ (G)$

The physical, the vital, and the mental levels are orientations in which patterns of perceiving, being, behaving are set in their ways. Each pattern has its purpose and its limitation and it can be argued that being able to learn from each orientation and yet being able to move beyond that, is the next logical step in any developmental model. The integral level hence, is about being able to leverage each of the patterns that naturally arise at the three preceding levels at will, and about further, being able to integrate these and arrive at new ways of perceiving and being.

Mathematically such behavior may be represented as being an integrative function ($\int x)$ where 'x' is the ability to move between the patterns emanating from G, e, π, at will, represented by $G, \bar{e}, \pi$. Integral-level (I) modulation (mod) of uniqueness (*Sig*) can hence be represented by the following equation-segment:

$$I: \; Sig * mod \left(\int G, \bar{e}, \pi \right)$$

The condition of overcoming any fixed and limiting patterns is the prerequisite for the emergence of 'Force' or for entering into the force-level. At this level the uniqueness behind the particular development being considered can emerge in its purity and become a truly creative dynamic. This aspect of creativity that is in a sense not bound by circumstance may be represented by the constant 'c', the speed of light in a vacuum, which is an upper limit of the layer that systems practically operate in. This is also likely the level at which N-entanglement is overtly active. Force-level (F) modulation (mod) of uniqueness (*Sig*) can hence be represented by the following equation-segment:

$$F: \; Sig * mod \, (c)$$

Once the signature of an organization arises and continues to exercise itself in its purity, it achieves contextual-mastery (C) and is able to exercise itself as though the context it is acting in, that can vary in scale and complexity, were all of the same substance as itself. This is likely the level at which K-entanglement is active. This equality may be represented by the integrative function '$\int = 1$'. The equation-segment that notates this contextual-level (C) modulation (mod) applied to organizational uniqueness (*Sig*) is hence:

$$C: \; Sig * mod \left(\int = 1 \right)$$

Piecing all the equation-segments together the equation for the emergence of uniqueness (Sig_E), where 'X' can be any of the discussed modulations at the respective development-model levels (P, V, M, I, F, C), is hence summarized by Equation 2.5.3:

$$Sig_E = X \begin{vmatrix} C\text{:}Sig * mod \left(\int = 1 \right) \\ F\text{:}Sig \ mod \ (c) \\ I\text{:} Sig \ mod \left(\int G, \bar{e}, \pi \right) \\ M\text{:}Sig * mod \ (G) \\ V\text{:}Sig * mod \ (e) \\ P\text{:}Sig * mod \ (\pi) \end{vmatrix}$$

Eq 2.5.3: Emergence of Uniqueness

Chapter 2.6: Inherent Dynamics of Any System

So far the inherent innovation that exists at the system level and summarized by the nature of a point and its attendant $'\infty$- entanglement' has been considered. Further, how this deep fount of innovation is present everywhere and how sets with their attendant 'K-entanglement' that make more practical the range of creative forces available in each of the four components of a point have also been considered. These architectural forces further define the possibility inherent in any system. Leveraging these sets of forces by virtue of 'N-entanglement' an equation for the uniqueness of an organization, regardless of scale, was arrived at.

In some sense the precipitation of innovation from the barely perceptible nature of the ubiquitous point, to how this reveals a play of forces, to how organizations take their seed and grow from these forces, giving insight too into the dynamics of entanglement and superposition in Space and Time, has been traced.

It is useful to now turn full-circle to return to the initial orientations that allowed so much to be suggested about the nature of innovation in the first place. It is useful to look deeper into the nature of the physical, the vital, the mental, and the integral, and to derive equations that in effect will provide further insight into the dynamics of innovation inherent in these orientations.

Hence, starting with the physical, an equation, Equation 2.6.1, is summarized as:

$$Physical = \begin{bmatrix} M_3 \to System_{Pr} \\ (\uparrow F \to I) \\ M_2 \to S_{Systemi_{Pr}} \\ (\uparrow Sig \to F) \\ M_1 \to Sig_P \\ (\uparrow > P_P) \\ U \to Physical_U \end{bmatrix} TC \to Physical_T$$

$$Where \begin{bmatrix} Physical_U \ni [inertia, lethargy, status\ quo,...] \\ Physical_T \ni [adaptability, durability, strength,...] \end{bmatrix}$$

Eq 2.6.1: Inherent Dynamism in Physical

Essentially this equation is laying out the conditions of moving from the untransformed or negative physical state represented by $Physical_U$ to the transformed or positive physical state represented by $Physical_T$.

The first matrix should be read from the bottom to the top:

$$\begin{bmatrix} M_3 \rightarrow System_{Pr} \\ (\uparrow F \rightarrow I) \\ M_2 \rightarrow S_{System_{Pr}} \\ (\uparrow Sig \rightarrow F) \\ M_1 \rightarrow Sig_P \\ (\uparrow > P_p) \\ U \rightarrow Physical_U \end{bmatrix}$$

Hence, at the bottom is the starting point '$U \rightarrow Physical_U$' which identifies the default or untransformed (U) level of the physical. The next row up, $(\uparrow > P_p)$, states that when the patterns of the untransformed physical (P_P) have been overcome (>), movement to the next level ($\uparrow$) is facilitated. Breaking through to the next level, $M_1 \rightarrow Sig_P$, allows its dynamics to become active. Hence, the signature or uniqueness of the physical (Sig_P) becomes active at meta-level 1 (M_1). As this signature becomes more like a Force ($Sig \rightarrow F$), the conditions for breakthrough ($\uparrow$) to the next level are achieved. This next level is referred to as meta-level 2 (M_2), and indicates that the architectural forces represented by the set of system-presence ($S_{System_{Pr}}$) have become more consciously active. When this Force becomes Integral ($F \rightarrow I$) then the conditions for breakthrough ($\uparrow$) to the next level are achieved. The next level is notated as M_3 for meta-level 3, and the dynamics here indicate that the equation for system-presence becomes active. Becoming active basically means that the respective meta-level dynamic begins to act at the once 'untransformed' level (U) further modifying it. Modification or transformation began when M_1 became active. Transformation is accelerated when M_2 becomes active, and even further accelerated when M_3 becomes active.

The notion of meta-layers is being explored by contemporary physicists and Erwin Laszlo in his book Self-Actualizing Cosmos (Laszlo, 2014) summarizes some of these developments: "Physicists describe the domain that underlies and embeds the particles, fields, and forces of the universe variously as quantum vacuum, physical spacetime, nuether, zero-point field, grand-unified field, cosmic plenum, or string-net liquid." Note that 'nuether' refers to a sub-quantum level of reality (Pearson, 1997). Laszlo goes on to describe a discovery by Nika Arkani-Hamed of Princeton's Institute of Advanced Study, of a geometrical object, the amplituhedron (Arkani-Hamid et al, 2012), which is not in space-time but governs space-time so that it "appears that spatio-temporal phenomena are the consequence of geometrical relationships in a deeper dimension of physical reality". A deeper dimension of a physical layer suggests synonymity with a meta-layer.

The rate of the transformation can be better envisioned when considering action of the Transformation Circle, or TC. The TC can be thought of as 4 concentric circles, with M_3 at the center. M_3 is surrounded by M_2, which is surrounded by M_1. The outer circle is U. If TC is considered to be a clock, than at time 't = 0', the physical' can be thought of as being entirely in U. The clock starts ticking only when some initial patterns P_P are overcome ($>^{P}P)$. From this point on as time proceeds the conditions for breakthrough become riper, and a sinusoidal wave begins to integrate more of the concentric circles together. The sinusoid wave (sin) is itself modulated by an euler function, e^x, where 'x' is determined by the strength to overcome patterns ($\uparrow$) which will likely vary over time but will likely tend to be positive once the clock has started ticking because of the joy experienced with progressive movement. Being that the limit is the outer boundary of the concentric circles, there is further modulation by π until the 4 concentric circles have been integrated. TC, hence, may be represented by Equation 2.6.2:

$$TC \equiv (> P_P) \to \text{mod} (\sin, e^x, \pi)$$

Eq 2.6.2: Transformation Circle

Hence, the initial nature of the physical that may be characterized by the set comprising of elements such as, lethargy, acceptance of the status quo, amongst other such elements, is represented by:

$(Physical_U \ni [inertia, lethargy, status\ quo, ...])$

This transforms into a physical more characterized by elements such as adaptability, durability, strength, and so on. That is:

$(Physical_T \ni [adaptability, durability, strength, ...])$

57

This transformation represents the inherent innovation-dynamic within the Physical.

Such transformation as discussed in the previous chapters will implicitly involve the mechanisms of superposition and entanglement as emergence takes place.

Similarly, the equation for the 'Vital', Equation 2.6.3, also shows the built-in transformation that represents the innovation-dynamic within the vital:

$$Vital = \begin{bmatrix} M_3 \to System_P \\ (\uparrow F \to I) \\ M_2 \to S_{System_P} \\ (\uparrow Sig \to F) \\ M_1 \to Sig_V \\ (\uparrow > P_V) \\ U \to Vital_U \end{bmatrix} TC \to Vital_T,$$

$$Where \begin{bmatrix} Vital_U \ni [aggression, self\ centeredness, exploitation,...] \\ Vital_T \ni [energy, support, adventure, enthusiasm,...] \end{bmatrix}$$

Eq 2.6.3: Inherent Dynamism of Vital

The equation for the 'Mental', Equation 2.6.4, is similarly summarized as:

$$Mental = \begin{bmatrix} M_3 \to System_S \\ (\uparrow F \to I) \\ M_2 \to S_{System_S} \\ (\uparrow Sig \to F) \\ M_1 \to Sig_M \\ (\uparrow > P_M) \\ U \to Mental_U \end{bmatrix} TC \to Mental_T$$

$$Where \begin{bmatrix} Mental_U \ni [fixation, fundamentalism, fragmentation,...] \\ Mental_T \ni [understanding, imagination, inspiration,...] \end{bmatrix}$$

Eq 2.6.4: Inherent Dynamism of Mental

The equation for the 'Integral', Equation 2.6.5, is similarly summarized as:

$$Integral = \begin{bmatrix} M_3 \rightarrow System_N \\ (\uparrow F \rightarrow I) \\ M_2 \rightarrow S_{System_N} \\ (\uparrow Sig \rightarrow F) \\ M_1 \rightarrow Sig_I \\ (\uparrow > P_I) \\ U \rightarrow Integral_U \end{bmatrix} TC \rightarrow Integral_T$$

$$Where \begin{bmatrix} Integral_U \ni [possession, usurpation, hidden \ agendas, ...] \\ Integral_T \ni [appreciation, shift \ POV, MPV, synthesis, ...] \end{bmatrix}$$

Eq 2.6.5: Inherent Dynamism of Integral

The preceding equations can be generalized by Equation 2.6.6:

$$Innovation_{orientation-x} = \begin{bmatrix} M_3 \rightarrow System_X \\ (\uparrow F \rightarrow I) \\ M_2 \rightarrow S_{System_X} \\ (\uparrow Sig \rightarrow F) \\ M_1 \rightarrow Sig_x \\ (\uparrow > P_x) \\ U \rightarrow x_U \end{bmatrix} TC \rightarrow x_T \,, where \begin{bmatrix} x_U \ni [...] \\ x_T \ni [...] \end{bmatrix}$$

Eq 2.6.6: Generalized Equation of Innovation

In this generalized equation, $Innovation_{orientation-x}$, refers to the inherent innovation within a specific orientation. Orientation refers to the physical, the vital, the mental, or the integral.

Further, the notion of a core-matrix can be summarized by the following equation, Equation 2.6.7:

$$Core\_matrix = \begin{bmatrix} M_3 \rightarrow System_X \\ (\uparrow F \rightarrow I) \\ M_2 \rightarrow S_{System_X} \\ (\uparrow Sig \rightarrow F) \\ M_1 \rightarrow Sig_x \\ (\uparrow > P_x) \\ U \rightarrow x_U \end{bmatrix}$$

Eq 2.6.7: Core Matrix

One of the corollaries of the Generalized Equation of Innovation is that if the source of innovation is more influenced by a meta-level there will be simultaneity of innovation or emergence that becomes apparent at U. This a direct result of N-, K-, and ∞- entanglement. The higher the meta-level, hence, the more likely that this simultaneity will be wider spread. In his book Where Good Ideas Come From: The Natural History of Innovation (Johnson, 2010) Johnson states: "A brilliant idea occurs to a scientist or inventor somewhere in the world, and he goes public with his remarkable finding, only to discover that three other minds had independently come up with the same idea in the past year." He refers to an essay "Are Inventions Inevitable" (Ogburn & Thomas, 1922) which uncovered 148 instances of similar yet independent innovation, most of them occurring within the same decade. Some examples include sunspots that were simultaneously discovered in 1611 by four scientists in four different countries, and the law of conservation of energy that was formulated separately four times in the late 1840s, amongst numerous other examples.

What should become apparent as a result of this discussion is that there is a uniformity of phenomenon, regardless of scale or level of granularity. In other words, phenomenon such as entanglement and superposition, thought to be solely true of the quantum-levels are in fact true at any level of consideration from the micro to the macro.

Chapter 2.7: Qualified Determinism

Leveraging off the previous chapter a mathematical notion of qualified determinism is explored here - Dynamic Interaction (DI) that has a 'vertical' and a 'horizontal' component. The vertical component is designated as DI_V and the horizontal component as DI_H. Several equations to capture the inherent dynamism that exists in each orientation or state have already been derived in Chapter 2.6. These included equations for the dynamism in the physical, the vital, the mental, and the integral. The derived equations propose a model to give insight into how innovation occurs by changing the fundamental states that an organization is subject to. Several scientists, such as (Prigogine, 1977) and others, are proposing that a system can bifurcate in unpredictable ways to create an emergent property that cannot be predicted. DI is going to propose that in fact there is a 'qualified determinism' as opposed to randomness that occurs.

DI is going to propose an alternative to the paradigm of statistics and probability thought to govern quantum objects.

This qualified determinism is the result of the relative strengths of the levels within the core-matrix as summarized by Equation 2.6.7 and reproduced here for convenience:

$$\begin{bmatrix} M_3 \to System_X \\ (\uparrow F \to I) \\ M_2 \to S_{System_X} \\ (\uparrow Sig \to F) \\ M_1 \to Sig_x \\ (\uparrow > P_{x)} \\ U \to x_U \end{bmatrix}$$

The application of the vertical component of the new function being proposed, DI_V, to this core matrix will yield the nature or 'strength' of the state (x) or orientation under consideration. If the untransformed or U layer is strongest, implying that the habitual patterns that keep an organization locked into its untransformed way of operation are still very active, then the nature of the output of DI_V, notated by x-state, will be x_U. If the habitual patterns have been overcome then the strength of the x-state increases since it is the dynamics of M_1 or Sig_x that are now active. In this case the x-state will be Sig_x. If the unique 'signature' has become a 'force', then the conditions for activation of M_2 have been put in place and the x-state will be even higher, S_{System_X}. The

61

architectural forces active in M_2 are by definition more powerful than Sig_x that is a derivation of a set of such architectural forces. If the 'force' so acting becomes impersonal so that an organizational ego-state is overcome, then the x-state will have the most strength and is characterized by $System_x$ active at M_3. Hence, DI_v applied to a core-matrix will yield the 'strength' in terms of the x-dynamic that is active. This is illustrated by the following equation, Equation 2.7.1, which can be considered to be a deductive proof in the context of this model:

$$DI_V \begin{bmatrix} M_3 \rightarrow System_X \\ (\uparrow F \rightarrow I) \\ M_2 \rightarrow S_{System_X} \\ (\uparrow Sig \rightarrow F) \\ M_1 \rightarrow Sig_x \\ (\uparrow > P_{x)} \\ U \rightarrow x_U \end{bmatrix} \Rightarrow$$

$$\underset{\text{x-state}}{\in} \left(x_U, Sig_x, S_{System_X}, System_X \right)$$

$Where:\ Strength(System_X) > Strength(S_{System_X}) > Strength(Sig_x) > Strength(x_U)$

Eq 2.7.1: Illustrating Action of Dynamic Interaction – vertical component

What is to be noted here is that while the action of DI_v yields a relative strength and therefore a 'single' value for the core- or x-matrix under consideration yet each x-matrix in itself could have an infinite number of possibilities. This should be clear in looking at how x_U, Sig_x, S_{System_X}, and $System_X$, were initially defined.

Hence, taking the example where x = physical:

$Physical_U \ni [inertia, lethargy, status\ quo, ...]$

As can be seen $Physical_U$, defined in Chapter 2.6, is practically speaking already an infinite set with qualities similar to the ones specified.

Similarly, Sig_{Pr}, defined in Chapter 2.4, also has an infinite variation:

$$Sig_{Pr} = Xa + Y\overline{b}_{0-n} \quad where \begin{bmatrix} X \in [S_{System_{Pr}}] \\ Y \in [S_{System_{Pr}}, S_{System_P}, S_{System_K}, S_{System_N}] \\ a, b \ are \ integers; a > b \end{bmatrix}$$

$S_{System_{Pr}}$, defined in Chapter 2.3, is also an infinite set with forces of the nature specified in the following equation:

$$S_{System_{Pr}} \ni [Service, Perfection, Diligence, Perseverance, ...]$$

And recall that in Chapter 2.2, $System_{Pr}$ has been defined as:

$$System_{Pr} \equiv TS_{0 \to N} \begin{bmatrix} Presence \\ \downarrow \\ Opportunity \end{bmatrix} \begin{bmatrix} P_L & \to & V_L \\ V_L & \to & M_L \end{bmatrix} \& \ Container \begin{bmatrix} System_P \\ System_K \\ System_N \end{bmatrix}$$

So in essence DI_V is really giving us a summary assessment of the 'level' of the x-matrix under consideration with all its infinite potentiality. An example will follow shortly.

The other component of DI, as suggested earlier in this chapter, is the horizontal component, DI_H. Just as DI_V yields a summary assessment of the level that an x-matrix is operating at, similarly DI_H yields a summary assessment of the direction that a system or organization under consideration is going to continue its development in, considering the physical, the vital, the mental, and the integral orientations to be the choices.

Assuming that any organization or system is inherently unique, as this mathematical model proposes, and as the phenomena of entanglement proves, and assuming that the infinite sets of x_U and S_{System_X} applied across the physical, vital, mental, and integral orientations respectively will account for any state that an organization can experience, then at a certain point in time any organization under consideration is going to have a direction-bias in one of the possible physical, vital, mental, or integral directions. Hence, DI_H will yield the summary direction that is going to lead an organization into its future given the current states active in it.

This summary direction is going to be yielded by considering the relative strengths of the separate core x-matrices – the physical, the vital, the mental, and the integral. The assumption is that there will be one core-matrix that will be stronger than the others.

Hence, as an example, first applying DI_V across all four x-matrices may, for example, yield the following results, with the strongest level within each x-matrix highlighted and bolded:

$$
\begin{bmatrix} System_{Pr} \\ S_{System_{Pr}} \\ Sig_P \\ Physical_U \end{bmatrix}
\begin{bmatrix} System_P \\ S_{System_P} \\ Sig_V \\ Vital_U \end{bmatrix}
\begin{bmatrix} System_K \\ S_{System_K} \\ Sig_M \\ Mental_U \end{bmatrix}
\begin{bmatrix} System_N \\ S_{System_N} \\ Sig_I \\ Integral_U \end{bmatrix}
$$

Since by definition the strength of $System_x$ is greater than S_{System_x}, which is greater than Sig_x, which is greater than x_U, applying DI_H, as in Equation 2.7.2, across these x-matrices, as in the example following it will then yield the strongest direction, which in this example is the Physical:

$$
DI_H \left(\begin{bmatrix} System_{Pr} \\ S_{System_{Pr}} \\ Sig_P \\ Physical_U \end{bmatrix}
\begin{bmatrix} System_P \\ S_{System_P} \\ Sig_V \\ Vital_U \end{bmatrix}
\begin{bmatrix} System_K \\ S_{System_K} \\ Sig_M \\ Mental_U \end{bmatrix}
\begin{bmatrix} System_N \\ S_{System_N} \\ Sig_I \\ Integral_U \end{bmatrix} \right) = Orientation_{Strongest}
$$

Eq 2.7.2: Illustrating Action of Dynamic Interaction - horizontal component

Example:

$$
DI_H \left(\begin{bmatrix} System_{Pr} \\ S_{System_{Pr}} \\ Sig_P \\ Physical_U \end{bmatrix}
\begin{bmatrix} System_P \\ S_{System_P} \\ Sig_V \\ Vital_U \end{bmatrix}
\begin{bmatrix} System_K \\ S_{System_K} \\ Sig_M \\ Mental_U \end{bmatrix}
\begin{bmatrix} System_N \\ S_{System_N} \\ Sig_I \\ Integral_U \end{bmatrix} \right) = Physical
$$

Hence, DI function will yield the following organizational direction, as in Equation 2.7.3, where 'x_matrix' is used interchangeably with 'orientation':

$$Org\_Dir = DI \left| \begin{array}{cc} \begin{bmatrix} M_3 \to System_{Pr} \\ (\uparrow F \to I) \\ M_2 \to S_{System_{Pr}} \\ (\uparrow Sig \to F) \\ M_1 \to Sig_P \\ (\uparrow > P_{P)} \\ U \to Physical_U \end{bmatrix} & \begin{bmatrix} M_3 \to System_{P} \\ (\uparrow F \to I) \\ M_2 \to S_{System_{P}} \\ (\uparrow Sig \to F) \\ M_1 \to Sig_V \\ (\uparrow > P_{V)} \\ U \to Vital_U \end{bmatrix} \\ \begin{bmatrix} M_3 \to System_{S} \\ (\uparrow F \to I) \\ M_2 \to S_{System_{S}} \\ (\uparrow Sig \to F) \\ M_1 \to Sig_M \\ (\uparrow > P_{M)} \\ U \to Mental_U \end{bmatrix} & \begin{bmatrix} M_3 \to System_{N} \\ (\uparrow F \to I) \\ M_2 \to S_{System_{N}} \\ (\uparrow Sig \to F) \\ M_1 \to Sig_I \\ (\uparrow > P_{I)} \\ U \to Integral_U \end{bmatrix} \end{array} \right| \to$$

$$x\_matrix_{strongest} @ level_{strongest}$$

Eq 2.7.3: Organizational Direction

Generalizing, as in Equation 2.7.4, where Org_Dir is organizational direction:

$$Org\_Dir = DI \left| \begin{bmatrix} M_3 \to System_{X} \\ (\uparrow F \to I) \\ M_2 \to S_{System_{X}} \\ (\uparrow Sig \to F) \\ M_1 \to Sig_x \\ (\uparrow > P_{P)} \\ U \to x_U \end{bmatrix} x = p, v, m, i \right| \to$$

$$x\_matrix_{strongest} @ level_{strongest}$$

Eq 2.7.4: Generalized Equation for Organizational Direction

Hence, this mathematical model is suggesting that any situation, rather than having a random outcome, has a 'qualified deterministic' outcome. In the introduction to his book "Where is Science Going?" (Planck, 1933), James Murphy points out that the reason Planck spent so much of his time giving lectures on causation was because of the trend of physicists at the time, which has continued to the modern day, to overthrowing the principle of causation following the development of quantum theory, which he felt was misplaced. "Planck would claim", he wrote, "and so would Einstein, that it is not the principle of causation itself which has broken down in modern physics, but rather the traditional formulation of it." Murphy also quotes James Jeans (Jeans, 1932) to suggest the issue associated with causation and determinism: "Einstein showed in 1917 that the theory founded by Planck appeared, at first sight at least, to entail consequences far more revolutionary than mere discontinuity", and here he is referring to the finding that radiant energy is not emitted in a continuous flow, but in integral quantities, or quanta, which can be expressed in integral numbers. Continuing: "It appeared to dethrone the law of causation from the position it had therefore held as guiding the course of the natural world. The old science had confidently proclaimed that nature could follow only one road, the road which was mapped out from the beginning of time to its end by the continuous chain of cause and effect; state A was inevitably succeeded by state B. So far the new science has only been able to say that state A may be followed by state B or C or D or by innumerable other states. It can, it is true, say that B is more likely than C, C than D, and so on; it can even specify the relative probabilities of B, C, and D. But, just because it has to speak in terms of probabilities, it cannot speak with certainty which state will follow which; this is a matter which lies on the knees of the gods – whatever gods there may be."

While under the apparent dynamics at the quantum level there may appear to be randomness and a dethroning of the principle of causation, the notion of a multiplicity of levels, each having its impact on the strength of an orientation and further on the consequent direction from a multiplicity of possible orientations, is being suggested here as determining the direction of any system, while still allowing infinite variation in the details that may define its. Hence, the positions of Planck and Einstein are vindicated when considering Equation 2.7.4.

Further, assuming any system where multiple elements are active, connected, interdependent, and emergent, it may be possible to understand, through application of calculus, as to which level is the source for change.

Hence, where N may be source of change, the rate of change of N will resolve into one of P_U, V_U, M_U, I_U, P_T, V_T, M_T, or I_T. This may be summarized by Equation 2.7.5, where y is either U or T:

$$\frac{dN}{dt} \rightarrow \begin{bmatrix} P_U & P_T \\ V_U & V_T \\ M_U & M_T \\ I_U & I_T \end{bmatrix} \rightarrow x_y, \text{ where } y \in (U, T)$$

Eq 2.7.5: Establishing the Nature of the Change

If T, implying that the action of one of the meta-levels has caused transformation, then application of one of the following integrals will determine which level is the likely source for change.

Hence, for M_1, if the integral of $\dfrac{\partial(x_U \to x_T)}{\partial t}$ across a limited area 'a' in the vicinity of the change, is greater than some threshold value $Threshold_{Signature}$, then the signature dynamics are likely the source of change. This is summarized by Equation 2.7.6:

$$\int_0^a \frac{\partial(x_U \to x_T)}{\partial t} > Threshold_{Signature}$$

Eq 2.7.6: Signature Dynamics as the Source of Change

For M_2, if the integral of $\dfrac{\partial(x_U \to x_T)}{\partial t}$ across a larger area 'b' extending beyond the vicinity of the change, is greater than some threshold value $Threshold_{Architectural\,Forces}$, then the architectural forces are likely the source of change. This is summarized by Equation 2.7.7:

$$\int_0^b \frac{\partial(x_U \to x_T)}{\partial t} > Threshold_{ArchitecturalForces}$$

Eq 2.7.7: Architectural Forces as the Source of Change

For M_3, if the double integral of $\dfrac{\partial(x_U \to x_T)}{\partial t}$ across the system specified by 'A', and across some time 't', is greater than some threshold value $Threshold_{System\,Property}$, then the system properties are likely the source of change. This is summarized by Equation 2.7.8:

$$\int_0^t \int_0^A \frac{\partial(x_U \to x_T)}{\partial t} > Threshold_{SystemProperty}$$

Eq 2.7.8: System Properties as the Source of Change

SECTION 3: THE MATHEMATICS OF QUANTIZATION & FOUNDATION OF HUMAN-QUANTUM COMPUTATIONAL MODELS

Building on the mathematical foundation laid out in Section 2, this section will further elaborate some mathematics in integrating quantum-levels with human-based dynamics.

Specifically this section focuses on the derivation of the Light-Space-Time Emergence equation leveraged in subsequent quantization analyses. The Light-Space-Time Emergence equation models the basis for quantization suggesting emergent reality for all phenomena from Light. Schrodinger's wave equation and Heisenberg's uncertainty principle are also interpreted from the point of view of the Light-based Interpretation of Quantum Mechanics to reinforce the notion that even when considered from these points of view the existence of multiple layers of light is feasible. The chapter on quantization of space, time, matter and gravity models how these phenomena are related to Light and subsequently also models how these fundamentals work together to impact material reality. Finally, deductively derived mathematical operators that may be representative of manipulating quantum phenomena as per this light-based interpretation will also be explored thus suggesting some basis for human-quantum computational models.

Chapters in this section will focus on:

- Equation for Light-Space-Time Emergence
- Speed of Light and Quanta
- Interpreting Schrodinger's Equation
- Interpreting Heisenberg's Uncertainty Principle
- A Deeper Look at Quantization of Space, Time, Matter, and Gravity
- Impact of Wavefunction and Quantization on Material Reality
- Mathematical Operators in Human-Quantum Computation

Chapter 3.1: Equation for Light-Space-Time Emergence

Equation 2.6.6, the generalized equation for innovation, can be restated as an evolving form true for all time, as in Equation 3.1.1:

$$Innovation_{orientation-x} = \left(\begin{bmatrix} M_3 \to System_X \\ (\uparrow F \to I) \\ M_2 \to S_{System_X} \\ (\uparrow Sig \to F) \\ M_1 \to Sig_x \\ (\uparrow > P_x) \\ U \to x_U \end{bmatrix} TC \to x_T \, , \, where \begin{bmatrix} x_U \ni [...] \\ x_T \ni [...] \end{bmatrix} \langle x_U \mid x_T \rangle \right)$$

Eq 3.1.1: Evolving Form of Generalized Equation of Innovation

The added notation of $\langle x_U \mid x_T \rangle$ implies that the output of the previous iteration of the equation of innovation, x_T, where the subscript T implies relatively-transformed, now becomes the input, x_U, for the next iteration of the equation, where U implies relatively-untransformed. Hence through time there is greater and greater transformation that pushes experienced reality to greater and greater levels of functional-richness.

But further, given that quanta is proposed to be a doorway to deeper worlds of Light, that in fact allow aspects of those worlds or layers to become active at the surface layer U, the question is when have those aspects become active in manifest time. The following timeline based on generally accepted models of universal history (Particle Data Group, 2015) suggests when. Note too that the subsequent sections of emergence of the electromagnetic spectrum, matter, and life, will explore in far grater detail some of the statements made in the following timeline:

- At time, $t \leq 0$ seconds, only M_3 is active, and then remains active for all $t < \infty$. Recall that M_3 represents the four-fold reality present in every point-instant.
- At time, $0 \geq t > \infty$, M_2 the set of architectural forces continually gets added to, thereby increasing the size of the sets of forces.
- At time, $t \geq 0$, space, time, energy, gravity, the first clear expression of the four-fold order, emerge and continue to evolve as ∞-entanglement, K-entanglement, N-entanglement, and the reality subsequent of seeds continues to complexify.

69

- At time, $0 > t \geq 10^{-36}$ seconds, the equation of Innovation, $Innovation_{orientation-x}$, is such that M_1 also becomes active. The activation of M_1 begins to result in unique expressions or signatures of the set of architectural forces, and in this case in the reality of the essentially ubiquitous electromagnetic-spectrum (EM Spectrum) as a vehicle of the four-fold order that expressed itself in all that existed and in all that unfolded from that point in time on.

- At time, $t \sim 10^{-10}$ seconds, fundamental particles emerge as an essential material basis of the four architectural forces that frame all further development. As in the case of the EM Spectrum this implies the activity of M_1, and then also of U.

- At time, $t \sim 3 \times 10^5$ years light atoms emerge, and at time $t \sim 10^9$ years heavier atoms in the stars emerge. These also imply the continued activity of M_1 and U.

- At time, $t \sim 13.8 \times 10^9$ years, a further clear expression of the same fourfold order as the bases of an even more complex organization, that of cellular life and all that is founded on it comes into being. This time-point will be represented by the notation $t \sim E_{Cell}$, where 'E' stands for emergence. This too implies the activity of M_1. Note that the sets of architectural forces specified by M_2 continue to increase the number of elements they comprise of as the complex interaction between the layers continues.

- At time $t > 13.8 \times 10^9$ years, human-beings, and more complex social organizations emerge. Here TC acts with an implicit direction of operation from U to M_3. This time-point will be represented by $t \sim E_{Human}$.

Based on the aforementioned description Equation 3.1.2 for Emergence true of any space-time scale may be generalized as the following:

$$Emergence_{space-time} = \begin{vmatrix} \begin{vmatrix} \begin{bmatrix} M_3 \rightarrow System_X \\ (\uparrow F \rightarrow I) \\ M_2 \rightarrow S_{System_X} \\ (\uparrow Sig \rightarrow F) \\ M_1 \rightarrow Sig_x \\ (\uparrow > P_x) \\ U \rightarrow x_U \end{bmatrix} Space \\ \begin{bmatrix} M_3 : -\infty \leq t \leq \infty \\ \downarrow \\ M_2 : 0 \geq t > \infty \\ \downarrow \\ M_1 : 0 > t > \infty \\ \downarrow \\ U \rightarrow \begin{array}{l} t \leq E_{Cell}; TC: M_3 \rightarrow U \\ t \sim E_{Human}; TC: U \rightarrow M_3 \end{array} \\ TC \rightarrow x_T, where \begin{bmatrix} x_U \ni [...] \\ x_T \ni [...] \end{bmatrix} \end{bmatrix} Time \end{vmatrix} \langle x_U | x_T \rangle \end{vmatrix}$$

Eq 3.1.2: Space-Time Emergence

An implication of this equation, brought out more explicitly through the elaboration of the 'Time' component, is that the layers U, M_1, M_2, and M_3 exist simultaneously. Adding the Light-Matrix derived in Chapter 2.1 enhances Equation 3.1.2 to the Light-Space-Time Emergence form as represented by Equation 3.1.3:

$Emergence_{light-space-time} =$

$$\begin{vmatrix} \begin{vmatrix} \begin{bmatrix} C_\infty : [Pr, Po, K, H] \\ (\downarrow R_{C_K} = f(R_{C_\infty})) \\ C_K : [S_{Pr}, S_{Po}, S_K, S_H] \\ (\downarrow R_{C_N} = f(R_{C_K})) \\ C_N : f(S_{Pr} \times S_{Po} \times S_K \times S_H) \\ (\downarrow R_{C_U} = f(R_{C_N})) \\ C_U : [P, V, M, C] \end{bmatrix} Light \\ \begin{bmatrix} M_3 : -\infty \leq t \leq \infty \\ \downarrow \\ M_2 : 0 \geq t > \infty \\ \downarrow \\ M_1 : 0 > t > \infty \\ \downarrow \\ U \rightarrow \begin{array}{l} t \leq E_{Cell}; TC: M_3 \rightarrow U \\ t \sim E_{Human}; TC: U \rightarrow M_3 \end{array} \end{bmatrix} Time \end{vmatrix} \begin{vmatrix} \begin{bmatrix} M_3 \rightarrow System_X \\ (\uparrow F \rightarrow I) \\ M_2 \rightarrow S_{System_X} \\ (\uparrow Sig \rightarrow F) \\ M_1 \rightarrow Sig_x \\ (\uparrow > P_x) \\ U \rightarrow x_U \end{bmatrix} Space \\ TC \rightarrow x_T \end{vmatrix} \langle x_U | x_T \rangle \end{vmatrix}$$

Eq 3.1.3: Light-Space-Time Emergence

In (3.1.3) there is a 1:1 mapping between the Light and Space matrices in that M_3 reflects the ever-present C_∞, M_2 reflects C_K, M_1 reflects C_N, and U reflects C_U. The Time matrix simply gives estimates at which time each of the layers became active.

Chapter 3.2: Speed of Light and Quanta

As discussed conceptually in Section 1 and also mathematically in Chapter 2.1, since c is finite and therefore there is past, present, and future implied by it, this implies that at U a point has to become quanta. This is implicit in the notion of finiteness. Since light takes a finite amount of time to get from A to B, a "unit" of light will require a finite time to traverse that. Quanta at the subatomic level can be thought of as related to this finite time and distance for a unit of light to be expressed.

Planck's discovery that energy at the subatomic level requires a minimum threshold 'quanta' to express itself therefore makes sense. It is to be noted though that Planck's treatment of quanta was more as a mathematical convenience that allowed the derivation of an equation that explained the curve of radiation wave-lengths at varying temperatures of a heated black-body (Isaacson, 2008). Einstein though postulated quanta as a fundamental property of light itself, rather than as something that arose in the interaction of light with matter as Planck thought. Einstein's theory produced a law of the photoelectric effect where the energy of emitted electrons would depend on the frequency of light. Einstein received the Nobel Prize for this discovery (Isaacson, 2008).

Summarizing, if c is the upper limit of the layer U, then it makes sense that the lower limit h (Planck's constant) should be inversely proportional to c. Hence:

$$h \propto \frac{1}{c}$$

This relationship is substantiated by combining two well-known equations: the first is the electromagnetic equation connecting speed of light with wavelength and frequency, and the second is Einstein's photoelectric equation connecting energy with frequency of light:

(1) $C = v\lambda$
(2) $E = hv$

Yields:

$$h = \frac{E\lambda}{C}$$

About h, H.A. Lorentz the Dutch scientist has commented in The Science of Nature (Lorentz, 1925): "We have now advanced so far that this constant not only furnishes the basis for explaining the intensity of radiation and the wavelength for which it represents a maximum, but also for interpreting the quantitative relations existing in several other cases among the many physical quantities it determines. I shall mention a few only, namely the specific heat of solids, the photo-chemical effects of light, the orbits of electrons in the atom, the wavelengths of the lines of the spectrum, the frequency of the Roentgen rays which are produced by the impact of electrons of given velocity, the velocity with which gas molecules can rotate, and also the distances between the particles which make up a crystal. It is no exaggeration to say that in our picture of nature nowadays it is the quantum conditions that hold matter together and prevent it from completely losing its energy by radiation."

So just as $\dot{c}$ sets up the past-present-future experience and reality of U, h suggests that this experience will take place in shells of matter. In the absence of the limit h, as pointed out by Lorentz, only radiation, and no matter would exist. This 'past-present-future-matter' dynamic reinforces the notion of the four-foldness implicit in the nature of light as already discussed in Sections 1 and 2.

The suggested variance in the speed of light by meta-layer may also throw some further light on the quantum realm. First, summarizing:

1. At U the speed of light in a vacuum, c_U, is finite at 186,000 miles/sec. This finiteness creates the reality and experience of past-present-future, and further a sense of fragmentation and separation. Further, assuming that c_U is a fundamental upper-limit at U, the inverse of it $\dfrac{1}{'c_U}$, must define some fundamental lower limit at U. This is indeed the case as Planck's constant, h, is proportional to this. 'h' allows for matter to be sustained, as it fundamentally limits the dispersion of energy as suggested by Lorentz.

2. At M_3, the speed of light, c_{M_3}, is suggested as being ∞ miles/sec. This allows a reality of 'oneness' and the possibility of a suggested fourfold-intelligence existing in every point-instant as already suggested.

3. As also already suggested the quantum world, here designated by Q, because it is at boundary of U, accesses and interrelates with the meta-levels. As such, the speed of light, c_Q, will appear as a hybrid as in the following figure. Note

though that it is really the speed of light at the native or resident layer that becomes active, and that this is simply being represented as c_Q for convenience:

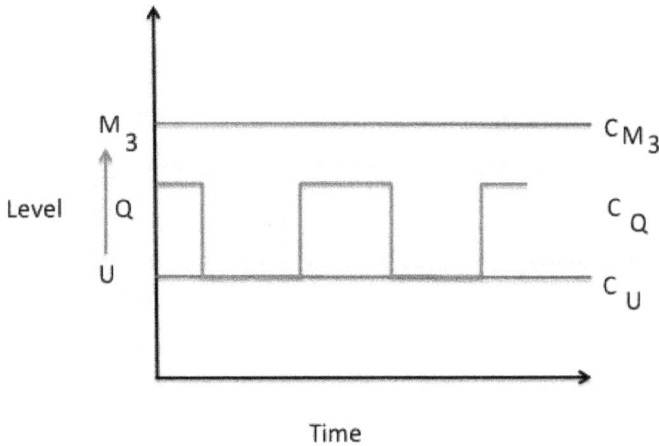

Figure 3.2.1 Speed of Light at Quantum Level

Note that research on the speed of light also indicates that it may go faster than c_U. While the speeds suggested currently through experimental research, and summarized below, may be only incrementally higher than c_U the notion that c_U can be exceeded appears to be put in place:

1. The Heisenberg Uncertainty Principle (to be discussed subsequently in Chapter 3.4) already suggests that photons can travel at any speed, even exceeding c_U, for short periods.
2. Notion of different space-time realities, also known as meta-levels in this treatise, suggests that light can travel differently in a layer different from the four-dimensional space-time that apparently defines our observable world (Hawking, 1988)
3. In his book QED Feynman (Feynman, 1985) says "...there is also an amplitude for light to go faster (or slower) than the conventional speed of light. You found out in the last lecture that light doesn't go only in straight lines; now, you find out that it doesn't go only at the speed of light! It may surprise you that there is an amplitude for a photon to go at speeds faster or slower than the conventional speed, c." In research conducted at Humboldt University (Chown, 1990), Scharnhorst has

75

made calculations using the theory of quantum electrodynamics to reveal the possible existence of "faster-than-light" photons. This is known as the Scharnhorst effect.

4. As reviewed in Chapter 2.1, Perkowitz makes the point that the theory of relativity does not disallow particles already moving at c or greater.

The point is that the reality at Q is going to be different than the reality at U. This should be apparent from considering the relation of c_X to the consequent reality as elaborated in Chapter 2.1. In Q the fundamental lower limit, h, which allows matter to sustain itself, is itself going to fluctuate. Hence, as X tends to M_3, c_X will tend to infinity, and h will become a fraction of itself. As it becomes a fraction of itself the quantization effect will be lowered, and matter will get dispersed more and more easily to in effect take on a wave-like, or field-like, or even block-like appearance as also suggested in Chapter 2.1.

Chapter 3.3: Schrodinger's Equation & Multiple Layers of Light

Schrodinger's equation, which seeks to model how a quantum state of a quantum system changes with time, or in other words seeks to model matter as a wave rather than as a particle (Stewart, 2012), is depicted in Equation 3.3.1:

$$i\frac{h}{2\pi}\frac{\partial}{\partial x}\psi = \hat{H}\psi$$

Eq 3.3.1: Schrodinger's Equation

ψ depicts a wave form and can be thought of as a probable cloud of possible states. $\hat{H}$ is the Hamiltonian operator which is a focusing function, and in its essence what the equation may be suggesting is that the way a wave form changes over time is equivalent to some expressible state of the possibilities inherent in the cloud of possible states.

But the cloud of possible states is another way of saying that behind the layer U, form is represented in another way than at U. If the existence of the meta-levels, and in this case, of Sig_x at M_1, is considered possible, then it is far more reasonable to admit that form is configured by function and the very dynamics of what may appear to have been random, as some interpretations of quantum phenomena suggest, may now appear to be far more logical. There is now more *context* to interpreting observation at the quantum level.

Interestingly Schrodinger himself had misgivings about the applicability of this equation that seemed to apply at the quantum level, to the macro-world (Stewart, 2012). To bring his misgivings to light he invented a thought experiment concerning a cat. This cat would be in a superposed state in a quantum black-box. A radioactive particle, a decaying-particle detector, and a flask of poison, were the other inhabitants of the black-box. At some point the particle will decay, be detected, and as in the thought-experiment at that point, triggered by the decaying particle, the poison in the flask would be released. The cat would then die. But in the meanwhile the cat would be in a superposed states of being both dead and alive. Only when the box was opened would the wave function collapse and a single definite state emerge. This thought experiment is also considered the origin of the notion of 'superposition'.

Schrodinger was hoping to highlight the absurdity of the application of having a cat in both a dead and alive state at the macro-level. Instead physicists found this thought experiment to be sensible, even at the macro-level, and began to generalize the quantum theory based on this. Hence, for example the idea of superposition at the physical level began to be thought of as real. It is interesting to note that in his lectures on Schrodinger's equation Feynman (Gottlieb, 2013) has stated: "Where did he get that [equation] from? Nowhere. It is not possible to derive it from anything you know. It came out of the mind of Schrödinger".

Considering Schrodinger's equation in the light of the discussion on Q in Chapter 3.2, 'i' is a complex number and suggests the interplay of two dimensions, one being real, and one being 'imaginary'. But the 'imaginary' dimension could be thought of as none other than the meta-levels implicit in the mathematical model in this treatise, and suggested to be real at Q. Further, $\frac{h}{2\pi}$ is in line with the suggestion also made in Chapter 3.2 that h will have to become a fraction of itself as c increases. Hence, the change in the wave function, $\frac{\partial}{\partial x}\psi$, is intimately related to i and $\frac{h}{2\pi}$, and perhaps more fully makes sense when considered in the context of i x $\frac{h}{2\pi}$ x $\frac{\partial}{\partial x}\psi$, which has to be the case when dealing with the integration of dynamics of multiple levels of light.

Further, the change in the wave function, $\frac{\partial}{\partial x}\psi$, is related to $\hat{H}\psi$, and suggests that there is some system "energy", represented by the Hamiltonian, $\hat{H}$, that when applied to the existing wave, ψ, will indicate how the wave will be expressed going forward.

But as discussed, at Q the dynamics of Sig_x become real, and in fact is a fundamental organizing principle for all organization at U, and starting at the space-dimension of h.

Chapter 3.4: Heisenberg's Uncertainty Principle & Multiple Layers of Light

In his book, The Little Book of String Theory, Princeton University's Gubser (Gubser, 2010) describes the effect on approaching absolute zero temperature on molecules. He takes the example of water molecules and relates that one cannot make the water molecules colder than absolute zero, -273.15 Celsius, because there is no more thermal energy to suck out at that temperature. However, quantum uncertainty, the phenomenon which relates the momentum and location of electrons in atoms necessitates that the water molecules will still vibrate. Gubser suggests this by considering Heisenberg's uncertainty relation, reproduced in Equation 3.4.1:

$$\Delta p \; X \; \Delta x \geq \frac{h}{4\pi}$$

Eq 3.4.1: Heisenberg's Uncertainty Relation

In Equation 3.4.1, Δp is the uncertainty in a particle's momentum, Δx is the uncertainty in the particle's location, and h is the Planck's constant. In frozen water crystals it is precisely known where the water molecules are, and therefore Δx is fairly small. This means that Δp has to be considerably larger, and therefore that the water molecules are still vibrating even though they are at absolute zero. This innate vibration, known as 'quantum zero-point' energy, expresses the phenomenon of quantum fluctuations.

The Planck's constant order of magnitude (10^{-34}) though, suggests the boundary between U and M_1 and the quantum fluctuations, the uncertainty relation, and the quantum zero-point energy could be an expression of the essential Signature function, Sig_x, that is posited as a key formative force behind organization at U. In this interpretation the thermal energy describes the essential energy at U, while the uncertainty relation may suggest the phenomenon of function-precipitation from other layers of Light, "physically" linking M_1 and U. In this case it may be suggested that integration of meta-levels with the surface level, I_U^M, is indicated by the uncertainty relation, as in Equation 3.4.2:

$$I_U^M \rightarrow \Delta p \; X \; \Delta x \; \geq \; \frac{h}{4\pi}$$

Eq 3.4.2: Integration of Levels (Leveraging Heisenberg's Uncertainty Relation)

But further, it may also be suggested that the uncertainty principle itself is only valid at U, and that too, because of the finiteness of c. This finiteness as already suggested implies h, which implies that if the position of a particle is going to be observed by shining light on it, the light has to have at least a quantum of energy. But to determine the position of a particle accurately, light of a shorter wavelength would have to be used (Hawking, 1988) which would have to have a minimum amount of energy, which in turn would interfere with the velocity and hence momentum of the particle. The uncertainty in measuring the momentum could therefore be thought of as a consequence of the finiteness of the speed of light, c.

If c_U were to approach c_Q though, which as suggested in Chapter 3.2 could be anywhere between c and ∞ miles per second, the quantum would be smaller and the uncertainty in measuring position or momentum would be reduced. At c_{M_3} there would be no uncertainty since light would accurately tell both position and momentum definitively.

Hence, the uncertainty principle may be further qualified, as in Equations 3.4.3, 3.4.4, and 3.4.5:

$$@C_U: \Delta p \; X \; \Delta x \; \geq \; \frac{h}{4\pi}$$

Eq 3.4.3: Uncertainty Principle at U

$$@C_Q: \Delta p \; X \; \Delta x \rightarrow 0$$

Eq 3.4.4: Uncertainty Principle at Q

$$@C_{M_3}: \Delta p \; X \; \Delta x = 0$$

Eq 3.4.5: Uncertainty Principle at M3

The notion of position and momentum becoming finite at U also may imply that space, time, and quanta are emergent rather than absolute properties, as also suggested in Section 1. This is also the conclusion of Arkani-Hamed of the

Institute of Advanced Studies in the following thought experiment (Wolchover, 2013):

'Locality says that particles interact at points in space-time. But suppose you want to inspect space-time very closely. Probing smaller and smaller distance scales requires ever higher energies, but at a certain scale, called the Planck length, the picture gets blurry: So much energy must be concentrated into such a small region that the energy collapses the region into a black hole, making it impossible to inspect. "There's no way of measuring space and time separations once they are smaller than the Planck length," said Arkani-Hamed. "So we imagine space-time is a continuous thing, but because it's impossible to talk sharply about that thing, then that suggests it must not be fundamental — it must be emergent."

Unitarity says the quantum mechanical probabilities of all possible outcomes of a particle interaction must sum to one. To prove it, one would have to observe the same interaction over and over and count the frequencies of the different outcomes. Doing this to perfect accuracy would require an infinite number of observations using an infinitely large measuring apparatus, but the latter would again cause gravitational collapse into a black hole. In finite regions of the universe unitarity can therefore only be approximately known.'

Chapter 3.5: A Deeper Look at Quantization of Space, Time, Matter, and Gravity

The Light-Space-Time Emergence equation (3.1.3) derived in Chapter 3.1, and reproduced here for convenience, offers some insight into the process of quantization. Reproducing the equation:

$$Emergence_{light-space-time} = \begin{vmatrix} \begin{vmatrix} C_{\infty}:[Pr, Po, K, H] \\ \left(\downarrow R_{C_K} = f(R_{C_{\infty}})\right) \\ C_K: [S_{Pr}, S_{Po}, S_K, S_H] \\ \left(\downarrow R_{C_N} = f(R_{C_K})\right) \\ C_N: f(S_{Pr} \times S_{Po} \times S_K \times S_H) \\ \left(\downarrow R_{C_U} = f(R_{C_N})\right) \\ C_U: [P,V,M,C] \\ M_3: -\infty \leq t \leq \infty \\ \downarrow \\ M_2: 0 \geq t > \infty \\ \downarrow \\ M_1: 0 > t > \infty \\ \downarrow \\ U \rightarrow \begin{matrix} t \leq E_{Cell}; TC: M_3 \rightarrow U \\ t \sim E_{Human}; TC: U \rightarrow M_3 \end{matrix} \end{vmatrix}_{Light} \begin{vmatrix} M_3 \rightarrow System_X \\ (\uparrow F \rightarrow I) \\ M_2 \rightarrow S_{System_X} \\ (\uparrow Sig \rightarrow F) \\ M_1 \rightarrow Sig_x \\ (\uparrow > P_x) \\ U \rightarrow x_U \end{vmatrix}_{Space} \\ \qquad\qquad\qquad\qquad\qquad \begin{matrix} \\ TC \rightarrow x_T \end{matrix}_{Time} \end{vmatrix} \langle x_U | x_T \rangle$$

The Light-Matrix, the top left-hand matrix, suggests that there are particular kinds of quantization that could occur. Along the vertical realm these can be thought of as inter-relating one layer of the matrix with the previous layers. Hence the kind of quantization that is relevant to the physical layer, U, would relate ^{C}U with ^{C}x where x $\in$ (N, K, ∞), one of the possible speeds of light as per this mathematical treatise. This quantization can be represented by ^{h}U, where h stands for Planck's constant. Note that in this model there are several fundamental quantization possible that may inter-relate a specific light-layer with others above it. For the sake of simplicity it will be assumed that ^{h}U is the quanta experienced in the inter-relation between U and the collective-set of light-layers behind or above U.

In looking at the Light-Matrix it is also clear that each subsequent layer below the top layer, describing the reality set up by an infinite speed, is emergent. Hence space, time, matter, and gravity, that arise when Light slows down, can be thought of as emergent phenomena. The emergence itself can be thought of as a function of the Light-Matrix. At layer U, and as discussed in Chapter 1.2, Space is related to Light's property of Knowledge, Time is related to Light's property of Power, Matter or Energy is related to Light's property of Presence, and Gravity is related to Light's Property of Harmony or Nurturing. Each of these will emerge in a particular way and it is possible that there are multiple h_U's.

Hence, the Planck's constant for matter or energy, which we are already familiar with from past discoveries in science, can be depicted as h_{UPr}, since it is related to Presence (Pr). But similarly, quantization for space, time, and gravity could potentially be governed by other similar constants, referred to as h_{UK}, h_{UP}, and h_{UH}, and related to Knowledge (K), Power (P), and Harmony (H), respectively. Since the relationship between these constants, in absolute terms, is uncertain, this can be represented by the equality-inequality as depicted by Equation 3.5.1:

$$h_{UPr} \lesseqgtr h_{UK} \lesseqgtr h_{UP} \lesseqgtr h_{UH}$$

Eq 3.5.1: Equality-Inequality Relationship Between Different 'Planck' Constants

Further, the "quantization-window", that is, the window that quanta provides to layers of light behind U, as it were, potentially allows the precipitation of, or inter-relation with, or creation of a cohesive and compelling meta-function or signature as modeled by the C_N/M_1 layer. This quantization-window is positioned as being key in allowing a phenomenon of "quantum-certainty" to occur. Quantum certainty can be thought of as allowing space-time-energy-gravity quantization to occur and is described in detail in the book Quantum Certainty (Malik, 2017e), part of the 6-book Cosmology of Light series (Malik, 2017-2018). Subsequent Sections in this book will describe several examples of the activation of the multi-dimensional quantization that ensures change in the "material-fabric" at the quantum-levels, thereby setting up the logic by which material change will subsequently occur. Such quantization requires patterns to be overcome, and so long as these are, this will result in the phenomenon of quantum-certainty. The 'Time' matrix, in (3.1.3) reproduced above, indicates the general default direction under which such quantum-certainty can occur at Layer U. Generally at the pre-human level it may proceed more automatically by the dynamics of the system itself, as represented by the segment: $t \sim E_{Cell}; TC: M_3 \rightarrow U$. Beyond this level it will generally occur through the action of a cohesive will, as represented by the segment: $E_{Human}; TC: U \rightarrow M_3$. The opening of this quantization-window is modeled by the 'Space' matrix and will happen when habitual patterns are overcome so that 'will' or 'want' becomes cohesive.

83

Such inter-dependence between human want or will and quantum dynamics is the basis of Human-Quantum computational models.

The specific quantization that are occurring along the space, time, energy/matter, and gravity dimensions are modeled by the following equations, that are all based on the Sig_x equations derived in Chapter 2.4. As a reminder (2.4.5) is reproduced here for convenience:

$$Sig = Xa + Y\bar{b}_{0-n}$$

$$where: \begin{bmatrix} X \in [S_{System_{Pr}}, S_{System_P}, S_{System_K}, S_{System_N}] \\ Y \in [S_{System_{Pr}}, S_{System_P}, S_{System_K}, S_{System_N}] \\ a, b \ are \ integers; a > b \end{bmatrix}$$

The Structure of Space

The structure of Space, which holds the seeds of knowledge of all that will emerge, hence, is modeled in the following way as specified by Equation 3.5.2:

$$Space_{quantization} = h_{UK}(Xa + Y\bar{b}_{0-n})$$

$$where: \begin{bmatrix} X \in [S_{System_K}] \\ Y \in [S_{System_{Pr}}, S_{System_P}, S_{System_K}, S_{System_N}] \\ a, b \ are \ integers; a > b \end{bmatrix}$$

Eq 3.5.2: Space Quantization

In this model 'space' as an emergent property of Light, is structured by infinite seeds of knowledge. But because the emergence is taking place in a layer of reality generally itself structured by a finite speed of light, c, it has to be quantized. The quantization assures that the knowledge is not dissipated, but can accumulate to create seeds, and therefore, the structure of space itself. Such a structure of space appears consistent with recent models on the structure of space as explored in Rovelli's 'Reality Is Not What It Seems' (Rovelli, 2017). It is assumed that there is some kind of 'Planck's constant' in effect, that is modeled as being specific to the way knowledge or space may be quantized – hence, h_{UK}.

The Structure of Time

The structure of Time, which holds the inevitability of the seeds or knowledge emerging in a phased maturity, hence, is modeled in the following way as specified by Equation 3.5.3:

$$Time_{quantization} = h_{UP}(Xa + Y\bar{b}_{0-n})$$

$$where: \begin{bmatrix} X \in [S_{System_p}] \\ Y \in [S_{System_{Pr}}, S_{System_p}, S_{System_K}, S_{System_N}] \\ a, b \ are \ integers; a > b \end{bmatrix}$$

Eq 3.5.3: Time Quantization

In this model 'time' as an emergent property of Light, is structured by an inevitable process of maturity, which due to its inevitability, is related to power. But because the emergence is taking place in a layer of reality generally itself structured by a finite speed of light, c, it has to be quantized. The quantization assures that the power is not dissipated, but can accumulate to express phased maturity, and therefore, the structure of time itself. It is assumed that there is some kind of 'Planck's constant' in effect, that is modeled as being specific to the way power or time may be quantized – hence, h_{UP}.

Energy & Matter

Energy, which through a process of containment can result in Matter, hence, is modeled in the following way as specified by Equation 3.5.4:

$$Energy_{quantization} = h_{UPr}(Xa + Y\bar{b}_{0-n})$$

$$where: \begin{bmatrix} X \in [S_{System_{Pr}}] \\ Y \in [S_{System_{Pr}}, S_{System_p}, S_{System_K}, S_{System_N}] \\ a, b \ are \ integers; a > b \end{bmatrix}$$

Eq 3.5.4: Energy Quantization

In this model 'energy' as an emergent property of Light, results in the reality of matter. But because the emergence is taking place in a layer of reality generally itself structured by a finite speed of light, c, it has to be quantized. The quantization assures that the energy is not dissipated, but can accumulate to create matter. Planck's constant is referred to as - h_{UPr}.

Gravity

Gravity, which holds seemingly distinct objects in the layer of reality created by c together in a harmony, hence, is modeled in the following way as specified by Equation 3.5.5:

$$Gravity_{quantization} = h_{UH}\left(Xa + Y\bar{b}_{0-n}\right)$$

$$where: \begin{bmatrix} X \in [S_{System_N}] \\ Y \in [S_{System_{Pr}}, S_{System_P}, S_{System_K}, S_{System_N}] \\ a, b \ are \ integers; a > b \end{bmatrix}$$

Eq 3.5.5: Gravity Quantization

In this model 'gravity' as an emergent property of Light, results in a harmonious collectivity of seemingly independent objects. But because the emergence is taking place in a layer of reality generally itself structured by a finite speed of light, c, it has to be quantized. The quantization assures that the harmony is not dissipated, but can accumulate to express more and more complex collectivities on large-scale. It is assumed that there is some kind of 'Planck's constant' in effect, that is modeled as being specific to the way harmony or gravity may be quantized – hence, h_{UH}.

Chapter 3.6: Quantization Effect of Organizational Transformation on Material Reality

The Light-Space-Time Emergence equation (3.1.3) being iterative can be used to model emergence from simpler to more complex four-fold manifestations. In other words (3.1.3) suggests a computational approach to the development of the universe. This computational approach involves multiple layers of reality as suggested by (3.1.3) and is driven more by a process of qualified determinism than probability and statistics. This notion of qualified determinism is in contrast to the prevalent probabilistic view that has been erected as the cornerstone of quantum theory, **and to MIT's Seth Lloyd's Copenhagen-like** quantum superposition, probability-based computational approach to the development of the universe **as described in his book** 'Programming the Universe' (Lloyd, 2007).

An example of Light-Space-Time emergence model will be elaborated in Section 4 as an indication of the detailed process of emergence of matter, life, and complex organization discussed in Sections 5, 6, 7, and 8. However, in keeping with the current quantum-analysis paradigm of using wavefunction to explore outcomes at a particular space and time, this chapter will develop two general approaches, summarized by equations, to understanding the possible quantization effect of organizational transformation on material reality. While the natural line of development given the mathematical treatise explored in this book is to proceed with such an understanding using (3.1.3), the wavefunction depiction will be elaborated to suggest how (3.1.3) can be leveraged to provide further insight into a probability-based quantum-analytical approach.

(3.1.3) is reproduced below for convenience and a more concise form of it, the Simplified Version of Light-Space-Time Emergence, as in Equation 3.6.1 shall be utilized through the rest of this chapter.

$$\text{Emergence}_{light-space-time} = \left| \begin{array}{l} \underbrace{\begin{array}{l} C_\infty:[Pr,Po,K,H] \\ \left(\downarrow R_{C_K}=f(R_{C_\infty})\right) \\ C_K:[S_{Pr},S_{Po},S_K,S_H] \\ \left(\downarrow R_{C_N}=f(R_{C_K})\right) \\ C_N:f(S_{Pr}\times S_{Po}\times S_K\times S_H) \\ \left(\downarrow R_{C_U}=f(R_{C_N})\right) \\ C_U:[P,V,M,C] \end{array}}_{Light} \quad \underbrace{\begin{array}{l} M_3\to System_X \\ (\uparrow F\to I) \\ M_2\to S_{System_X} \\ (\uparrow Sig\to F) \\ M_1\to Sig_x \\ (\uparrow > P_x) \\ U\to x_U \end{array}}_{Space} \\[2em] \underbrace{\begin{array}{c} M_3:-\infty \le t \le \infty \\ \downarrow \\ M_2:0 \ge t > \infty \\ \downarrow \\ M_1:0 > t > \infty \\ \downarrow \\ U\to \begin{array}{l} t\le E_{Cell};TC:M_3\to U \\ t\sim E_{Human};TC:U\to M_3 \end{array} \end{array}}_{Time\ TC\to x_T} \end{array} \right|_{\langle x_U | x_T\rangle}$$

Simplifying each of the main matrices in (3.1.3), hence, yields Equation 3.6.1, the Simplified Version of the Light-Space-Time Emergence equation:

$$\text{Emergence}_{light-space-time} = \left| [L][S][T]TC\to x_T \right|_{\langle x_U | x_T\rangle}$$

Eq. 3.6.1: Simplified Version of Light-Space-Time Emergence

Quantization Effect on Material Reality Using Simplified Version of Light-Space-Time Emergence

As detailed in Chapter 2.6 on the inherent dynamics of any system, while relatively transformed organizations have opened to the active influence of meta-levels, relatively untransformed organizations remain under the influence of untransformed physical, vital, mental, and integral dynamics at U.

Equations 3.6.2 through 3.6.4 summarize three types of organizations using (3.6.1). These are the untransformed-organization with only untransformed bases active, the organization-in-transition with mixed bases, and the transformed organization with bases relatively transformed due to the active influence of meta-levels, respectively. Transformed and untransformed can be thought of as being computational outcomes of light-based algorithms of the type described in this book. Hence:

$$\text{Untransformed - organization} = \left\| \left| [L][S][T]TC\to x_T \right|_{\langle x_U | x_T\rangle} \right\|_U$$

Eq. 3.6.2: Untransformed-Organization

$$Organization - in - transition = \left\| [L][S][T]TC \rightarrow x_T \right|_{\langle x_U | x_T \rangle} \Big|_{U\&M_x}$$

Eq. 3.6.3: Organization-in-Transition

$$Transformed - organization = \left\| [L][S][T]TC \rightarrow x_T \right|_{\langle x_U | x_T \rangle} \Big|_{M_x}$$

Eq. 3.6.4: Transformed-Organization

Now the question is what is the level of quantization caused by an organization as specified by (3.6.2-4)? For only if quantization takes place will there be an effect on the material-fabric. Equation 3.6.5, Quantization Effect of Organization on Material-Fabric, summarizes how this may happen:

$$Impact\ on\ Materal - Fabric = \begin{bmatrix} \left\| [L][S][T]TC \rightarrow x_T \right|_{\langle x_U | x_T \rangle} \\ \times \\ \left(|mod(Z_Q)|_{Y>U} \right) \\ \ni \\ Z_Q, Z \in U\ (Space, Time, Energy, Gravity) \end{bmatrix}$$

Eq. 3.6.5: Quantization Effect of Organization on Material-Fabric

Line 1 in the matrix is simply (3.6.1) the Simplified Light-Space-Time Emergence equation. To understand the quantization that may occur, the organization resulting from the application of Line 1 is multiplied $(\times)$ by a modulation (mod) of a fourfold space-time-energy-gravity quantization (Z_Q), so long as Y>U, that is, the operative bases are relatively transformed and hence have an active influence greater than U. The fourfold quantization is specified by $(\ni Z_Q, Z \in U\ (Space, Time, Energy, Gravity))$ where $(\ni)$ the quantization being applied (Z) is each (U) of the elements of the set (Space, Time, Energy, Gravity).

Quantization Effect on Material Reality Using Wavefunction Form of Light-Space-Time Emergence

In Chapter 3.3 we were introduced to Schrodinger's wavefunction equation (3.3.1), reproduced below for convenience, which seeks to model matter as a wave rather than as a particle:

$$i\frac{h}{2\pi}\frac{\partial}{\partial x}\psi = \hat{H}\psi$$

This wavefunction equation deals with dynamics behind the surface, and particularly of as many as infinite potential superposed states that then collapse into a material possibility in time and space at U. Such superposition is another way of saying that there is wholeness behind the surface. In (3.6.1) such wholeness is represented by L, S, T, the Light, Space, and Time matrices respectively.

As explained by Neil Turok in his book The Universe Within (Turok, 2012), Euler's formula reproduced below, can be used to model many naturally occurring phenomena because of its sinusoidal oscillation between narrow bounds as x increases. Reproducing Euler's formula:

$$e^{ix} = \cos x + i\sin x$$

A key characteristic of this equation is that the sum of the squares of the ordinary and complex parts, on the right side of the equation, is one. In many contemporary interpretations of quantum theory this ensures that the probabilities for all possible outcomes add up to one. Dealing in probabilities become important when considering as many as infinite superposed states.

Further, in modeling material reality a modified notation of the Schrodinger wavefunction as interpreted by Feynman is leveraged. This version features the integral sign, $\int$, meaning that all terms to the right of it have to be summed up for all space and time till the moment when the wavefunction is required to be known.

Hence, combining (3.6.1) with Feynman's interpretation of the Schrodinger wavefunction, with the Euler formula yields the following equations, 3.6.6-11 for organizations:

$$\psi_{untransformed - organization} = \left| \int e^{i\int |[L][S][T]|} \right|_U$$

Eq. 3.6.6: Wavefunction for Untransformed-Organization

Note that in (3.6.6) the formulation specifying iteration, $\langle x_U \mid x_T \rangle$, becomes redundant since iteration is implied by the integral across space and time, and hence is removed. Further, (3.6.6) suggests that so long as the basis of an organization is untransformed, specified by the U following the vertical-brackets, the outcome is going to remain an untransformed organization, specified by $\psi_{untransformed-organization}$.

Further, as specified by Equation 3.6.7, the Probability-View of Untransformed-Organization, the bases of such organizations is going to be either the untransformed physical (^{P}U), the untransformed vital (^{V}U), the untransformed mental (^{M}U), or the untransformed integral (^{I}U):

$$\left| \psi_{untransformed-organization} \right|^2$$
$$= P_U^2 + V_U^2 + M_U^2 + I_U^2 = 1$$

Eq. 3.6.7: Probability-View for Untransformed-Organization

As specified by (3.6.7) the probability that any of these bases will be leveraged in the creation of an organization adds up to one. Note that the physical, the vital, the mental, and integral were discussed in Chapter 2.6 on Inherent Dynamics of a System.

Equation 3.6.8, Wavefunction for Organization-in-Transition, suggests the mixed bases for organizations, as specified by dynamics of both the untransformed (U) and the meta-levels (^{M}x), which therefore results in organization-in-transition.

$$\psi_{organization-in-transition} = \left| \int e^{i \int |[L][S][T]|} \right|_{U\&M_x}$$

Eq. 3.6.8: Wavefunction for Organization-in-Transition

As specified by Equation 3.6.9, Probability-View for Organization-in-Transition, the probability that any of the untransformed (U) and transformed (T) bases will be leveraged in the creation of an organization adds up to one:

$$\left| \psi_{organization-in-transition} \right|^2 = P_U^2 + P_T^2 + V_U^2 + V_T^2 + M_U^2 + M_T^2 + I_U^2 + I_T^2 = 1$$

Eq. 3.6.9: Probability View for Organization-in-Transition

Equation 3.6.10, Wavefunction for Transformed-Organization, suggests some transformed bases for an organization, as specified by dynamics of the meta-levels ($^M x$), which therefore results in a transformed-organization.

$$\psi_{transformed-organization} = \left| \int e^{i \int |[L][S][T]|} \right|_{M_x}$$

Eq. 3.6.10: Wavefunction for Transformed-Organization

As specified by Equation 3.6.11, Probability-View for Transformed-Organization, the probability that the transformed (T) bases will be leveraged in the creation of matter adds up to one:

$$\left| \psi_{transformed-organization} \right|^2 = P_T^2 + V_T^2 + M_T^2 + I_T^2 = 1$$

Eq. 3.6.11: Probability View for Transformed-Organization

Now the question is how would the very fabric of material existence be impacted based on the type of organization in effect and what might an equation that captured that look like? Equation 3.6.12, Quantization Effect of Organization on Material-Fabric (Wavefunction Form), is such an equation:

$$\psi_{impact-on-material-fabric} = \begin{bmatrix} \left(\left| \int e^{i \int |[L][S][T]|} \right|_{Y=(U,U\&M_x,M_x)} \right) \\ \times \\ \left(\left| mod(Z_Q) \right|_{Y>U} \right) \\ \ni \\ Z_Q, Z \in U \ (Space, Time, Energy, Gravity) \end{bmatrix}$$

Eq. 3.6.12: Quantization Effect of Organization on Material-Fabric (Wavefunction Form)

Taking each line in the equation separately: Line 1&2 from the top combines (3.6.6), (3.6.8), and (3.6.10) to essentially state that the wave-function for impact-on-fabric while governed by the Light-Space-Time matrices in (3.1.3) will vary based on the bases that are active, hence $Y = (U,U\&M_x,M_x)$, where Y equates to the family of bases that may be active: untransformed (U), a mix (U & $^M x$), or $^M x$. Since the primary part of Line 2 falls under the $\int$ it means that

92

the operation is iterative since dynamics are summed up for space and time to that point where the wavefunction is sought.

A quantization impact on the material-fabric of reality will only take effect, where 'taking effect' is specified by (mod) or being modulated by, if the bases of operation for the organization is more than just the untransformed (U) bases, hence Y>U. The impact on the material-fabric is the product (x) of Line 2 & 4, and the specific quantization ($^Z Q$) that will occur are each (^{U}) of the members ($\in$) of the set $(Space, Time, Energy, Gravity)$ so that the very fabric of reality is altered through corresponding space, time, energy, and gravity quantization.

In effect the altering of the fabric due to quantization, as discussed in the previous chapter sets up a field where a unique function or signature can have a material impact.

Chapter 3.7: Mathematical Operators in Human-Quantum Computational Models

Based on the mathematics developed in Section 2, twelve general 'mathematical operators' are arrived at deductively. These operators are the result of considering some of the dynamics of each of the levels, and integration of the multiple levels of the overall mathematical model. These are non exhaustive, but rather are indicative of the nature of operators that may arise in such a mathematical model.

These operators assume an added significance because they give additional insight into alternative quantum-related computational models. If, as the light-based interpretation of quantum dynamics suggests, superposition and entanglement are intimately associated with the dynamics of space and time then the case can be made that they are not just accessible by going down into the quantum realms. In fact subsequent chapters will highlight the place the human capacities such as will, thought, emotion, can have in mobilizing quantum realms. If this is true then really what may be possible is a genre of human-quantum computational models. Mathematical operators of the type described in this chapter may suggest some ways in which such human-quantum computational models may develop.

The following operators have to be considered in context to the Generalized Equation for Innovation, Equation 2.6.6, derived in Chapter 2.6. This equation suggests that any system has implicit in it the urge to transform the untransformed layer, U, by opening to the influence of the meta-layers, M_1, M_2, and M_3. There is interplay between the levels by virtue of habitual patterns that are broken. The breaking of habitual patterns interconnects different layers together, and of necessity, in this light-based quantum model and mathematics, this involved dynamics at the quantum levels. In so doing the very sources of innovation or computation are altered and the visible characteristics of systems are transformed by the action of these sources of innovation. Several sets have already been suggested that explore these sources of innovation - $S_{System_{Pr}}$, S_{System_P}, S_{System_K}, and S_{System_N}, and the resultant characteristics of systems - $Physical_T$, $Vital_T$, $Mental_T$, and $Integral_T$.

The following sections leverage various aspects of the already derived mathematical model to frame sets of mathematical operators along each of the four bases of Light that components or elements of systems will be subject to in their journey toward transformation.

Presence-Based Mathematical Operators

Considering Light's dimension of Presence, Equation 3.7.1, Presence-Based Mathematical Operators, summarizes a set of representative presence-based mathematical operators:

$$Presence\_based_{Mathematical\_operators} \ni [Fullness, Equality, Uniqueness, ...]$$

Eq. 3.7.1: Presence-Based Mathematical Operators

Fullness refers to the possibility that every single point in any time-space continuum is informed by Light's four-fold fullness. That is, the reality of ∞-entanglement is fully active. Hence, in equation form it may be suggested that Fullness is the union (U) of $System_{Pr}$, $System_P$, $System_K$, and $System_N$ as in Equation 3.7.2, Fullness:

$$Fullness = U \begin{bmatrix} System_{Pr} \\ System_P \\ System_K \\ System_N \end{bmatrix}$$

Eq. 3.7.2: Fullness

In general for any two points 'A' and 'B' it can be suggested that the Fullness behind A is the same as the Fullness behind B, as in Equation 3.7.3, Equivalence of Fullness. This suggestion may also be arrived at by considering Einstein's General Theory of Relativity (Einstein, 1995) that states: "All bodies of reference K, K1 etc. are equivalent for the description of natural phenomena (formulation of the general laws of nature), whatever may be their state of motion." If we shrink these bodies of reference or coordinate systems to infinitesimal dimensions, thus approaching a 'point', this suggests that there is equivalence in that the general laws of nature are equally valid at any two points. Hence, Equation 3.7.3:

$$Fullness_A \equiv Fullness_B$$

Eq. 3.7.3: Equivalence of Fullness

Equality refers to the possibility that since every point in a time-space continuum is always some expression of Light's four-fold fullness, hence every point is equal to any other point. This implies that all expressions and developments share a fundamental equality with all other points. Depicting a point 'a' by $Point_A$ and a point 'b' by $Point_B$, then an equation, Equation 3.7.4, Equality, would be:

Equality: $Point_A \equiv Point_B$

Eq. 3.7.4: Equality

Uniqueness refers to the possibility that any point is fundamentally unique. An easy way to envision this is to see that any point in a time-space continuum is a result of a unique time-space intersection. Hence, two points within a time-space continuum, A and B, can always be envisioned as having unique time and space coordinates as in Figure 3.7.1:

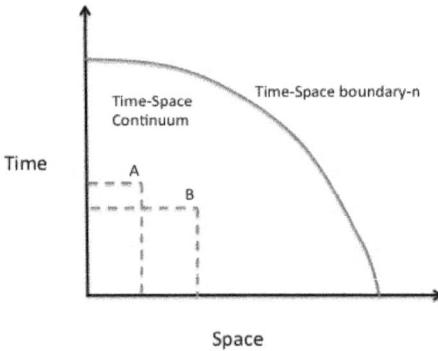

Figure 3.7.1: Uniqueness in Time-Space Continuum

But even for any organization, which can itself be considered a further development of a point or points through the action of time and space, the already proposed equation, (2.4.5) that governs organizational signatures may be restated as an equation for uniqueness, as in Equation 3.7.5:

$$Uniqueness:\ Sig = Xa + Yb_{0-n}^{-}\ \ where \begin{bmatrix} X \in [S_{System_{Pr}}, S_{System_{P}}, S_{System_{K}}, S_{System_{N}}] \\ Y \in [S_{System_{Pr}}, S_{System_{P}}, S_{System_{K}}, S_{System_{N}}] \\ a, b\ are\ integers; a > b \end{bmatrix}$$

Eq. 3.7.5: Uniqueness

Power-based mathematical operators pertain to the fount of dynamism that will tend to determine an organization's practical action. Mathematical operators related to 'power' give insight into how an organization will tend to meet circumstance and other organizations.

Representative operators include Direction, Fractal, and Intersection and in contrast to Knowledge-based operators discussed in the next sub-section, tend to create the more visceral and immediate reactions to circumstance. Knowledge-based operators on the other hand tend to create action more in line with long-term or strategic plans. This power-based set can be summarized by Equation 3.7.6:

$$Power\_based_{Mathematical\_operators} \ni [Direction, Fractal, Intersection...]$$

Eq. 3.7.6: Power-Based Mathematical Operators

Direction refers to the possibility that direction at any possible system bifurcation point is not random and not determined either. Rather it can be thought of as a qualified determinism, as introduced in Chapter 2.7, and is a function of applying DI_V and DI_H to the set of possibilities existing at a bifurcation point. As per the functioning of DI_V and DI_H, it is the strongest or most 'powerful' possibility that will tend to determine what will emerge as an organization meets circumstance and other organizations. Hence leveraging (2.7.4) creates Equation 3.7.7:

$$Direction: Org\_Dir = DI \left(\left. \left[\begin{matrix} M_3 \to System_X \\ (\uparrow F \to I) \\ M_2 \to S_{System_X} \\ (\uparrow Sig \to F) \\ M_1 \to Sig_x \\ (\uparrow > P_P) \\ U \to x_U \end{matrix} \right] \right|_{x = p, v, m, i} \right) \to$$

$$x\_matrix_{strongest} @ level_{strongest}$$

Eq. 3.7.7: Direction

Fractal refers to the possibility that as organizational complexity increases, the base orientation, orientation-x, where x could be physical, vital, mental, or integral, of an average organization at some level of complexity 'n', will tend to

97

determine the orientation of an organization at a level of complexity 'n+1'. Likewise the orientation of an organization at level of complexity 'n+1' will tend to determine the orientation of an organization at level of complexity 'n'. This is summarized by Equation 3.7.8:

$$Fractal:\ Orientation_x\ @\ Complexity_n \leftrightarrow Orientation_x\ @\ Complexity_{n+1}$$

Eq. 3.7.8: Fractal

Organizational complexity refers to an order of magnitude change as in from a person to a team, or from a team to a business unit, and so on, for example. In Nature such fractal arrangements abound in the way the human body is constructed to the very structure of galaxies (Briggs, 1992). This notion has been suggested to exist in complex behavioral systems as well as described in some detail in books such as The Fractal Organization (Hoverstadt, 2008), The Fractal Organization (Malik, 2015), and The Misbehavior of Markets (Mandelbrot, 2006). This notion relates to 'power' in that it is the patterns at one level that will tend to determine the patterns at another level, often preempting what may be a more logical choice based on reason. This kind of behavior has been suggested as causing cyclic fluctuations in stock and other markets where greed and fear often trump more intelligent and rational choices (Frost, 2005). Greed rises until fear sets in. Fear rises until greed sets in.

Intersection occurs when two organizations shift orientations to the next successive level due to the shock of interaction. Hence if an organization is at a physical orientation, it may be shifted to a vital orientation when intersection occurs, as in Equation 3.7.9, Intersection, where the function 'Next Element' extracts the next element from the Set S comprising the elements (physical, vital, mental, integral). Examples of such phenomena abound where failure to make a shift results in shock of conflict repeating itself endlessly. Such a process with applicability at multiple levels of complexity has been captured by the series of books on crucial conversations (Patterson, 2011).

$$Intersection:\ Organization\ Orientation \rightarrow Next\ Element\ (S)$$

Eq. 3.7.9: Intersection

Knowledge-Based Operators

Knowledge-based mathematical operators have to do with how organizations tend to develop by existing or toward increasing knowledge over time.

Representative knowledge properties discussed in this section include Alternative, Flowering, and Higher, which as suggested in the precious section are of a different nature than 'power' or dynamism-based properties that tend to determine an organization's visceral or immediate reaction to the market place. This knowledge-based set can be summarized by Equation 3.7.10:

$$Knowledge\_based_{Mathematical\_operators} \ni [Alternative, Flowering, Higher...]$$

Eq. 3.7.10: Knowledge-Based Mathematical Operators

Alternative refers to an alternative narrative that an organization will tend to embed itself in. These alternative narratives relate to the physical, the vital, the mental and the integral orientations. These narratives can easily become fixed and can strongly influence the entire internal and external orientation of an organization. The book 'The Fractal Organization: The Future of Enterprise' (Malik, 2015) suggests a theory of such narratives with their consequent effect on practical action. The alternative narratives are best described using the generalized equation (2.6.6) modified as Equation 3.7.11. Hence:

$$Alternative: Innovation_{orientation-x} = \begin{bmatrix} M_3 \to System_X \\ (\uparrow F \to I) \\ M_2 \to S_{System_X} \\ (\uparrow Sig \to F) \\ M_1 \to Sig_x \\ (\uparrow > P_x) \\ U \to x_U \end{bmatrix} TC \to x_T, where \begin{bmatrix} x_U \ni [...] \\ x_T \ni [...] \end{bmatrix}$$

Eq. 3.7.11: Alternative

Where, 'x' can be thought of as an element from the Set of Orientations:

$$x \in (physical, vital, mental, integral)$$

'X' can be thought of as an element from the Set of System-level architectural forces. Hence:

$$X \in (Presence, Power, Knowledge, Nurturing)$$

Flowering refers to the possibility that any time-space boundary-n, depicted as TS_n (as in Fig. 3.7.1) will have more potential or possibility associated with it than a time-space boundary-(n-1). Putting this into equation format yields Equation 3.7.12:

Flowering: $Possibility_{TS_n} > Possibility_{TS_{n-1}}$

Eq. 3.7.12: Alternative

Higher refers to the possibility that over time the direction will always tend to move to a higher meta-level. This can be summarized by using the notion of the core-matrix (2.6.7) yielding Equation 3.7.13:

$$Higher: Upward \begin{bmatrix} M_3 \to System_X \\ (\uparrow F \to I) \\ M_2 \to S_{System_X} \\ (\uparrow Sig \to F) \\ M_1 \to Sig_x \\ (\uparrow > P_x) \\ U \to x_U \end{bmatrix}$$

Eq. 3.7.13: Higher

Note also that while many organizations are practically at the untransformed level for a long time there is still movement within that level that can generally be depicted as a change from a predominantly physical-orientation to a more mental-orientation. In general this shift in orientation is implied, as more and more fixed patterns are overcome: $\uparrow > P_x$, where P_x refers to patterns along an orientation 'x' where x is an element from the set: (physical, vital, mental, integral).

Nurturing or Harmony-Based Mathematical Operators

Nurturing-based mathematical operators have to do with the nature of relationship within systems. These relationships are posited as being of a nurturing nature and emanate from the notion of 'nurturing' as an organizing class. Representative mathematical-operators include Remember, Linking, and Relate. This nurturing-based set can be summarized by Equation 3.7.14:

$Nurturing\_based_{Mathematical\_operators} \ni [Remember, Linking, Relate...]$

Eq. 3.7.14: Nurturing-Based Mathematical Operators

Remember has to do with remembering that there is something in each organization that existed before the existence of any organization, and further, that there is something in each organization that exists in every other organization. This can be thought of going back to a time-space moment of zero, and subsequently of expanding into the Time-Space Continuum keeping that connection in mind. There is something in each organization that exists in every organization and highlights a special way to relate to the underlying system as in Figure 3.7.2:

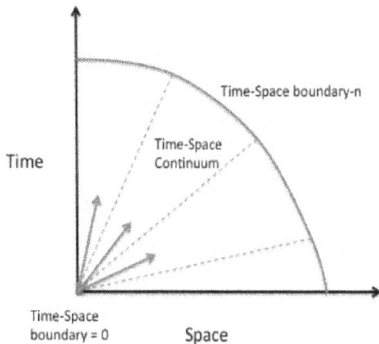

Figure 3.7.2: Remembrance in Time-Space Continuum

This may be depicted by Equation 3.7.15:

$$Remember:\ Ubiquity \begin{bmatrix} \overleftarrow{TS = 0} \\ \overrightarrow{TS > 0} \end{bmatrix}$$

Eq. 3.7.15: Remember

Where the condition of going back (←) before any organization existed, TS = 0, is invoked as all developments proceed (→), TS > 0, to create a sense of ubiquity. The sense of ubiquity is a remembrance. This notion may be akin to the concept that everything that is in the universe emanated from the Big Bang and that we are all recycled stardust (Swimme, 2001).

Link refers to the condition whereby in any time-space coordinate, irrespective of the level of untransformed reality (U), any present state can be consciously linked to the underlying ubiquitous system. This can be depicted by Equation 3.7.16, Link, where the conditions that usually need to be in place for a meta-layer to actively influence layer U disappear. There may be an attitude or receptiveness on the part of the element at U that allows such linking to take place. Hence:

$$Link: \begin{bmatrix} M_3 \rightarrow System_X \\ \uparrow\downarrow \\ M_2 \rightarrow S_{System_X} \\ \uparrow\downarrow \\ M_1 \rightarrow Sig_x \\ \uparrow\downarrow \\ U \dashrightarrow x_U \end{bmatrix}$$

Eq. 3.7.16: Link

Relate is a way to relate to the System so as to offer or surrender activities any kind of organization is involved in, to the System, and hence the Fullness or intelligence embedded in every point. This can be depicted by an Offer function, such that the first or relatively untransformed element in the function, depicted by x_U, is being offered to the second one, the union of the four-fold intelligence embedded in each point and depicted by the union function U[ ], as in Equation 3.7.17:

$$Relate: \quad Offer \left(x_U, \; \bigcup \begin{bmatrix} System_{Pr} \\ System_P \\ System_K \\ System_N \end{bmatrix} \right)$$

Eq. 3.7.17: Relate

Section 4: Overview of Real-Time Light-Space-Time Matrix Computations in Development of the Universe

If the mathematics presented in the previous sections is true, then the universe becomes the result of continuous, real-time Light-Matrix based computations implicitly involving quantum-levels. In fact if this is the case, then nothing can truly change unless superposition and entanglement are leveraged.

This section therefore briefly outlines how it may be that Light-Matrix based computations have created everything from the Big Bang to modern day Global Civilization, to whatever else may emerge.

Note that the sections following this will go into more detail to elaborate the process by which quantization incrementally changes material reality through alteration of the material-fabric.

Chapter 4.1: Application of Light-Space-Time Emergence Equation - Pre-Big-Bang Existence to an Evolving-Universe

The Light-Space-Time Emergence equation (3.1.3) reproduced below for convenience essentially provides an algorithm or structure by which Pre-Big-Bang existence proceeds through a reality of an ever-evolving universe.

$$Emergence_{light-space-time} = \begin{Vmatrix} \begin{Vmatrix} \begin{matrix} C_\infty:[Pr, Po, K, H] \\ (\downarrow R_{C_K} = f(R_{C_\infty})) \\ C_K: [S_{Pr}, S_{Po}, S_K, S_H] \\ (\downarrow R_{C_N} = f(R_{C_K})) \\ C_N: f(S_{Pr} \times S_{Po} \times S_K \times S_H) \\ (\downarrow R_{C_U} = f(R_{C_N})) \\ C_U: [P,V,M,C] \end{matrix} \end{Vmatrix}_{Light} \begin{Vmatrix} M_3 \to System_X \\ (\uparrow F \to I) \\ M_2 \to S_{System_X} \\ (\uparrow Sig \to F) \\ M_1 \to Sig_x \\ (\uparrow > P_x) \\ U \to x_U \end{Vmatrix}_{Space} \\ \begin{Vmatrix} \begin{matrix} M_3: -\infty \le t \le \infty \\ \downarrow \\ M_2: 0 \ge t > \infty \\ \downarrow \\ M_1: 0 > t > \infty \\ \downarrow \\ U \to \begin{matrix} t \le E_{Cell}; TC: M_3 \to U \\ t \sim E_{Human}; TC: U \to M_3 \end{matrix} \end{matrix} \end{Vmatrix}_{Time} TC \to x_T \end{Vmatrix} \langle x_U | x_T \rangle$$

Pre-Big-Bang Reality

Before the Big Bang it can be assumed, as per the discussion in Section 1, that only Light at ∞ existed. Hence (3.3.1) collapses to the highlighted portion only, to create a modified version, Equation 4.1.1, Pre-Big-Bang Reality, in which the rest of the Light, Space, and Time matrices are implicit:

$Pre - Big - Bang =$

$$\left|\begin{array}{l} \left|\begin{array}{c} C_{\infty}:[Pr, Po, K, H] \\ \left(\downarrow R_{C_K} = f(R_{C_\infty})\right) \\ C_K: [S_{Pr}, S_{Po}, S_K, S_H] \\ \left(\downarrow R_{C_N} = f(R_{C_K})\right) \\ C_N: f(S_{Pr} \times S_{Po} \times S_K \times S_H) \\ \left(\downarrow R_{C_U} = f(R_{C_N})\right) \\ C_U: [P, V, M, C] \end{array}\right| Light \quad \left|\begin{array}{c} M_3 \to System_X \\ (\uparrow F \to I) \\ M_2 \to S_{System_X} \\ (\uparrow Sig \to F) \\ M_1 \to Sig_x \\ (\uparrow > P_x) \\ U \to x_U \end{array}\right| Space \\[2em] \left|\begin{array}{c} M_3 : -\infty \le t \le \infty \\ \downarrow \\ M_2 : 0 \ge t > \infty \\ \downarrow \\ M_1 : 0 > t > \infty \\ \downarrow \\ U \to \begin{array}{l} t \le E_{Cell}; TC: M_3 \to U \\ t \sim E_{Human}; TC: U \to M_3 \end{array} \end{array}\right| Time \; TC \to x_T \end{array}\right| \langle x_U | x_T \rangle$$

Eq 4.1.1: Pre-Big-Bang Reality

The Big-Bang

Assuming Light then projects itself from its state of ∞ to that of c instantaneously, this creates the Big Bang and yields Equation 4.1.2, Big-Bang, where the highlighted portions only are operative. Note that essentially in this state the multiple quantization discussed in Section 3 becomes active through the projection of Light into slower moving layers of itself.

$Big - Bang =$

$$\left[\begin{array}{l} C_\infty:[Pr, Po, K, H] \\ \left(\downarrow R_{C_K} = f(R_{C_\infty})\right) \\ C_K: [S_{Pr}, S_{Po}, S_K, S_H] \\ \left(\downarrow R_{C_N} = f(R_{C_K})\right) \\ C_N: f(S_{Pr} \times S_{Po} \times S_K \times S_H) \\ \left(\downarrow R_{C_U} = f(R_{C_N})\right) \\ C_U: [P, V, M, C] \end{array}\right]_{Light} \left[\begin{array}{l} M_3 \to System_X \\ (\uparrow F \to I) \\ M_2 \to S_{System_X} \\ (\uparrow Sig \to F) \\ M_1 \to Sig_x \\ (\uparrow > P_x) \\ U \to x_U \end{array}\right]_{Space}$$

$$\left[\begin{array}{l} M_3 : -\infty \le t \le \infty \\ \qquad \downarrow \\ M_2 : 0 \ge t > \infty \\ \qquad \downarrow \\ M_1 : 0 > t > \infty \\ \qquad \downarrow \\ U \to \begin{array}{l} t \le E_{Cell}; TC: M_3 \to U \\ t \sim E_{Human}; TC: U \to M_3 \end{array} \end{array}\right]_{Time} \quad TC \to x_T \qquad \langle x_U \mid x_T \rangle$$

Eq 4.1.2: Big-Bang

Space-Time-Energy-Gravity Emergence

As discussed in Chapter 3.5 Space, Time, Energy, and Gravity are emergent and are the result of Layer U being created with the 'settling' down of Light to a speed c. Hence Equation 4.1.3, Space-Time-Energy-Gravity Emergence is synonymous with (4.1.2):

Space – Time – Energy – Gravity Emergence =

$$\left[\begin{array}{l} C_\infty:[Pr, Po, K, H] \\ \left(\downarrow R_{C_K} = f(R_{C_\infty})\right) \\ C_K: [S_{Pr}, S_{Po}, S_K, S_H] \\ \left(\downarrow R_{C_N} = f(R_{C_K})\right) \\ C_N: f(S_{Pr} \times S_{Po} \times S_K \times S_H) \\ \left(\downarrow R_{C_U} = f(R_{C_N})\right) \\ C_U: [P, V, M, C] \end{array}\right]_{Light} \left[\begin{array}{l} M_3 \to System_X \\ (\uparrow F \to I) \\ M_2 \to S_{System_X} \\ (\uparrow Sig \to F) \\ M_1 \to Sig_x \\ (\uparrow > P_x) \\ U \to x_U \end{array}\right]_{Space}$$

$$\left[\begin{array}{l} M_3 : -\infty \le t \le \infty \\ \qquad \downarrow \\ M_2 : 0 \ge t > \infty \\ \qquad \downarrow \\ M_1 : 0 > t > \infty \\ \qquad \downarrow \\ U \to \begin{array}{l} t \le E_{Cell}; TC: M_3 \to U \\ t \sim E_{Human}; TC: U \to M_3 \end{array} \end{array}\right]_{Time} \quad TC \to x_T \qquad \langle x_U \mid x_T \rangle$$

Eq 4.1.3: Space-Time-Energy-Gravity Emergence

It is with (4.1.3) that the emergence through matter, life, and mega-organization can proceed. For the Light-Matrix allows quantization to occur which means that the means for affecting the fabric of material existence, referred to as material-fabric, is now in place.

As a reminder in the preceding adaptations of (3.1.3) the notion of evolution is captured by $\langle x_U \mid x_T \rangle$, which signifies x_T becoming the new starting point for x_U as the process of adaptability continues.

Hence, if we trace the potential in Light working from the basis of an initial electromagnetic spectrum to the current basis of a possible sustainable global civilization, and beyond, some of the iterations the equation may have gone through can be envisioned as follows in Equation 4.1.4 – 11. Note that these are only snapshots and always there will be a vaster number of iterations that will be required.

Quantum Particle Emergence

In Equation 4.1.4, the starting point, x_U, is 'EM Spectrum', and the ending point, x_T, is 'Quantum Particles':

Quantum Particle Emergence =

$$
\left[
\begin{array}{l}
\left|
\begin{array}{l}
C_\infty : [Pr, Po, K, H] \\
\left(\downarrow R_{C_K} = f(R_{C_\infty}) \right) \\
C_K : [S_{Pr}, S_{Po}, S_K, S_H] \\
\left(\downarrow R_{C_N} = f(R_{C_K}) \right) \\
C_N : f(S_{Pr} \, x \, S_{Po} \, x \, S_K \, x \, S_H) \\
\left(\downarrow R_{C_U} = f(R_{C_N}) \right) \\
C_U : [P, V, M, C]
\end{array}
\right|_{Light}
\left|
\begin{array}{l}
M_3 \rightarrow System_X \\
(\uparrow F \rightarrow I) \\
M_2 \rightarrow S_{System_X} \\
(\uparrow Sig \rightarrow F) \\
M_1 \rightarrow Sig_x \\
(\uparrow > P_x) \\
U \rightarrow x_{EM \; Spectrum}
\end{array}
\right|_{Space} \\[2em]
\left|
\begin{array}{l}
M_3 : -\infty \leq t \leq \infty \\
\qquad \downarrow \\
M_2 : 0 \geq t > \infty \\
\qquad \downarrow \\
M_1 : 0 > t > \infty \\
\qquad \downarrow \\
U \rightarrow \begin{array}{l} t \leq E_{Cell}; TC : M_3 \rightarrow U \\ t \sim E_{Human}; TC : U \rightarrow M_3 \end{array}
\end{array}
\right|_{Time} \; TC \rightarrow x_{Quantum \; particles}
\end{array}
\right] \langle x_U \mid x_T \rangle
$$

Eq 4.1.4: Quantum Particle Emergence

As discussed in detail in Section 2, Mathematical Foundations For A Light-Based Interpretation of Quantum Phenomena, the process of evolution proceeds through a constant breaking of 'patterns'. Section 3, and in particular Chapters 3.5 and 3.6 lays out the mathematical process of quantization effecting material reality. Section 5 will discuss the computation leading to the emergence of the electromagnetic spectrum, while Section 6 will discuss the computations leading to the emergence of matter, including quantum particles.

Atoms Emergence

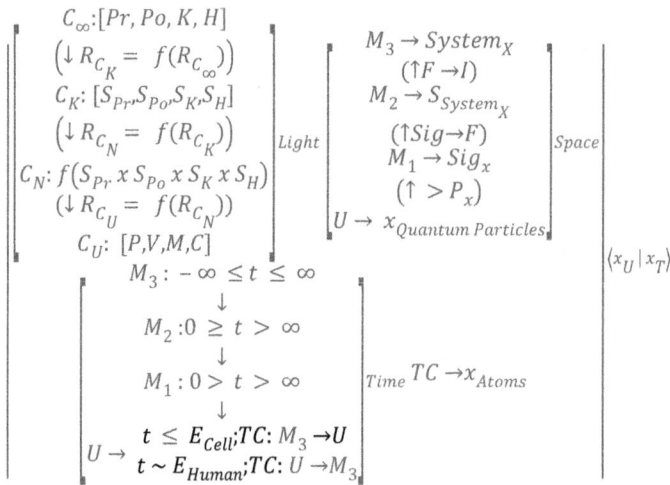

In Equation 4.1.5, Atoms Emergence, the starting point, x_U, is 'Quantum Particles', and the ending point, x_T, is 'Atoms':

Atoms Emergence =

$$
\left| \begin{array}{c}
\left| \begin{array}{c}
\left| \begin{array}{c}
C_\infty:[Pr, Po, K, H] \\
\left(\downarrow R_{C_K} = f(R_{C_\infty}) \right) \\
C_K: [S_{Pr}, S_{Po}, S_K, S_H] \\
\left(\downarrow R_{C_N} = f(R_{C_K}) \right) \\
C_N: f(S_{Pr} \times S_{Po} \times S_K \times S_H) \\
\left(\downarrow R_{C_U} = f(R_{C_N}) \right) \\
C_U: [P,V,M,C]
\end{array} \right| Light \quad
\left| \begin{array}{c}
M_3 \rightarrow System_X \\
(\uparrow F \rightarrow I) \\
M_2 \rightarrow S_{System_X} \\
(\uparrow Sig \rightarrow F) \\
M_1 \rightarrow Sig_x \\
(\uparrow > P_x) \\
U \rightarrow x_{Quantum\ Particles}
\end{array} \right| Space \\
\left| \begin{array}{c}
M_3: -\infty \leq t \leq \infty \\
\downarrow \\
M_2: 0 \geq t > \infty \\
\downarrow \\
M_1: 0 > t > \infty \\
\downarrow \\
U \rightarrow \begin{array}{l} t \leq E_{Cell}; TC: M_3 \rightarrow U \\ t \sim E_{Human}; TC: U \rightarrow M_3 \end{array}
\end{array} \right| Time \quad TC \rightarrow x_{Atoms}
\end{array} \right| \langle x_U | x_T \rangle
$$

Eq 4.1.5: Atoms Emergence

The computation leading to the emergence of atoms is discussed in greater detail in Chapter 6.3.

Molecules Emergence

In Equation 4.1.6, Molecules Emergence, the starting point, x_U, is 'Atoms', and the ending point, x_T, is 'Molecules':

Molecules Emergence =

$$
\begin{vmatrix}
\begin{vmatrix}
\begin{bmatrix}
C_\infty:[Pr, Po, K, H] \\
\left(\downarrow R_{C_K} = f(R_{C_\infty})\right) \\
C_K: [S_{Pr}, S_{Po}, S_K, S_H] \\
\left(\downarrow R_{C_N} = f(R_{C_K})\right) \\
C_N: f(S_{Pr} \times S_{Po} \times S_K \times S_H) \\
\left(\downarrow R_{C_U} = f(R_{C_N})\right) \\
C_U: [P, V, M, C]
\end{bmatrix} Light
&
\begin{bmatrix}
M_3 \to System_X \\
(\uparrow F \to I) \\
M_2 \to S_{System_X} \\
(\uparrow Sig \to F) \\
M_1 \to Sig_x \\
(\uparrow > P_x) \\
U \to x_{Atoms}
\end{bmatrix} Space \\
\begin{bmatrix}
M_3 : -\infty \leq t \leq \infty \\
\downarrow \\
M_2 : 0 \geq t > \infty \\
\downarrow \\
M_1 : 0 > t > \infty \\
\downarrow \\
U \to \begin{array}{l} t \leq E_{Cell}; TC: M_3 \to U \\ t \sim E_{Human}; TC: U \to M_3 \end{array}
\end{bmatrix} Time & TC \to x_{Molecules}
\end{vmatrix}
& \langle x_U | x_T \rangle
\end{vmatrix}
$$

Eq 4.1.6: Molecules Emergence

Cells Emergence

In Equation 4.1.7, Cells Emergence, the starting point, x_U, is 'Molecules', and the ending point, x_T, is 'Cells':

Cells Emergence =

$$\left| \begin{array}{l} C_\infty:[Pr, Po, K, H] \\ \left(\downarrow R_{C_K} = f(R_{C_\infty})\right) \\ C_K: [S_{Pr}, S_{Po}, S_K, S_H] \\ \left(\downarrow R_{C_N} = f(R_{C_K})\right) \\ C_N: f\left(S_{Pr} \times S_{Po} \times S_K \times S_H\right) \\ \left(\downarrow R_{C_U} = f(R_{C_N})\right) \\ C_U: [P,V,M,C] \end{array} \right| \; {}_{Light} \; \left| \begin{array}{l} M_3 \to System_X \\ (\uparrow F \to I) \\ M_2 \to S_{System_X} \\ (\uparrow Sig \to F) \\ M_1 \to Sig_x \\ (\uparrow > P_x) \\ U \to x_{Molecules} \end{array} \right| \; {}_{Space}$$

$$\left| \begin{array}{c} M_3: -\infty \leq t \leq \infty \\ \downarrow \\ M_2: 0 \geq t > \infty \\ \downarrow \\ M_1: 0 > t > \infty \\ \downarrow \\ U \to \begin{array}{l} t \leq E_{Cell}; TC: M_3 \to U \\ t \sim E_{Human}; TC: U \to M_3 \end{array} \end{array} \right| \; {}_{Time} \; TC \to x_{Cells} \qquad \langle x_U \,|\, x_T \rangle$$

Eq 4.1.7: Cells Emergence

The computation leading to the emergence of cells is discussed in greater detail in Chapter 7.1.

Humans Emergence

In Equation 4.1.8, Humans Emergence, the starting point, x_U, is 'Cells', and the ending point, x_T, is 'Humans'. Clearly there will be a vast number of iterations before humans emerge:

Humans Emergence =

$$\left| \begin{array}{l} C_\infty:[Pr, Po, K, H] \\ \left(\downarrow R_{C_K} = f(R_{C_\infty})\right) \\ C_K: [S_{Pr}, S_{Po}, S_K, S_H] \\ \left(\downarrow R_{C_N} = f(R_{C_K})\right) \\ C_N: f(S_{Pr} \times S_{Po} \times S_K \times S_H) \\ \left(\downarrow R_{C_U} = f(R_{C_N})\right) \\ \quad\quad C_U: [P,V,M,C] \end{array} \right|_{Light} \left| \begin{array}{c} M_3 \rightarrow System_X \\ (\uparrow F \rightarrow I) \\ M_2 \rightarrow S_{System_X} \\ (\uparrow Sig \rightarrow F) \\ M_1 \rightarrow Sig_x \\ (\uparrow > P_x) \\ U \rightarrow x_{Cells} \end{array} \right|_{Space} \langle x_U \,|\, x_T \rangle$$

$$\left| \begin{array}{l} M_3 : -\infty \leq t \leq \infty \\ \quad\quad \downarrow \\ M_2 : 0 \geq t > \infty \\ \quad\quad \downarrow \\ M_1 : 0 > t > \infty \\ \quad\quad \downarrow \\ U \rightarrow \begin{array}{l} t \leq E_{Cell}; TC: M_3 \rightarrow U \\ t \sim E_{Human}; TC: U \rightarrow M_3 \end{array} \end{array} \right|_{Time} TC \rightarrow x_{Humans}$$

Eq 4.1.8: Humans Emergence

If we follow the lines of emergence as will be described in Sections 7 and 8 then (4.1.8) will be the starting point for potentially different trajectories of development. In one trajectory 'Humans' will evolve 'Basic Capacities of Self' and then 'Truer Individuality'. In another line of development 'Humans' will be the starting point for 'Mega-Organization' culminating in 'Sustainable Global Civilization'. The gist of these can be represented by Equations 4.1.9 and 4.1.10 respectively.

Truer Individuality Emergence

Hence, Equation 4.1.9, Truer Individuality Emergence, the starting point, x_U, is 'Humans', and the ending point, x_T, is 'Truer Individuality'. Clearly there will be a vast number of iterations before 'Truer Individuality' emerges, that will contain 'Basic Capacities of Self' as a milestone along the way:

Truer Individuality Emergence =

$$
U \to \left\|
\begin{array}{l}
\left|
\begin{array}{l}
C_\infty : [Pr, Po, K, H] \\
\left(\downarrow R_{C_K} = f(R_{C_\infty})\right) \\
C_K : [S_{Pr}, S_{Po}, S_K, S_H] \\
\left(\downarrow R_{C_N} = f(R_{C_K})\right) \\
C_N : f(S_{Pr} \times S_{Po} \times S_K \times S_H) \\
\left(\downarrow R_{C_U} = f(R_{C_N})\right) \\
C_U : [P, V, M, C] \\
\end{array}
\right| \; Light \\
\left|
\begin{array}{l}
M_3 : -\infty \le t \le \infty \\
\quad \downarrow \\
M_2 : 0 \ge t > \infty \\
\quad \downarrow \\
M_1 : 0 > t > \infty \\
\quad \downarrow \\
t \le E_{Cell} ; TC: M_3 \to U \\
t \sim E_{Human} ; TC: U \to M_3 \\
\end{array}
\right| \; Time
\end{array}
\right\|
\quad
\left|
\begin{array}{l}
M_3 \to System_X \\
(\uparrow F \to I) \\
M_2 \to S_{System_X} \\
(\uparrow Sig \to F) \\
M_1 \to Sig_x \\
(\uparrow > P_x) \\
U \to x_{Humans} \\
\end{array}
\right| \; Space
\quad
\langle x_U \mid x_T \rangle
$$

$$TC \to x_{Truer\ Individuality}$$

Eq 4.1.9: Truer Individuality Emergence

Sustainable Global Civilization Emergence

In equation 4.1.10, Sustainable Global Civilization Emergence, the starting point, x_U, is 'Humans', and the ending point, x_T, is 'Sustainable Global Civilization'. Clearly there will be a vast number of iterations before 'Sustainable Global Civilization' emerges, that will contain 'Mega-Organization' as a milestone along the way:

Sustainable Global Civilization Emergence =

$$\left| \begin{matrix} \begin{matrix} C_\infty:[Pr,Po,K,H] \\ \left(\downarrow R_{C_K} = f(R_{C_\infty})\right) \\ C_K:[S_{Pr},S_{Po},S_K,S_H] \\ \left(\downarrow R_{C_N} = f(R_{C_K})\right) \\ C_N: f\left(S_{Pr} \times S_{Po} \times S_K \times S_H\right) \\ \left(\downarrow R_{C_U} = f(R_{C_N})\right) \\ C_U:[P,V,M,C] \end{matrix} \quad Light \quad \begin{matrix} M_3 \to System_X \\ (\uparrow F \to I) \\ M_2 \to S_{System_X} \\ (\uparrow Sig \to F) \\ M_1 \to Sig_x \\ (\uparrow > P_x) \\ U \to x_{Humans} \end{matrix} \quad Space \\ \\ \begin{matrix} M_3: -\infty \le t \le \infty \\ \downarrow \\ M_2: 0 \ge t > \infty \\ \downarrow \\ M_1: 0 > t > \infty \\ \downarrow \\ U \to \begin{matrix} t \le E_{Cell}; TC: M_3 \to U \\ t \sim E_{Human}; TC: U \to M_3 \end{matrix} \end{matrix} \quad Time \quad ^{TC} \to x_{Sust.\,Global\,Civilization} \end{matrix} \right| \quad \langle x_U | x_T \rangle$$

Eq 4.1.10: Sustainable Global Civilization Emergence

The computation leading to the emergence of a sustainable global civilization is discussed in greater detail in Chapter 8.3.

Super-Matter Emergence

In Equation 4.1.11, Super-Matter Emergence, the starting point, x_U, is 'Established Matter', and the ending point, x_T, is 'Super-Matter'. As per the mathematical model here all the preceding transformations change the reality of matter through subtle quantization that allows space, time, energy, and gravity to operate differently. The following equation is therefore one approximation in stating how super-matter may emerge:

Super – Matter Emergence =

$$
\left| \left| \begin{array}{l} \left[\begin{array}{c} C_\infty: [Pr, Po, K, H] \\ \left(\downarrow R_{C_K} = f(R_{C_\infty}) \right) \\ C_K: [S_{Pr}, S_{Po}, S_K, S_H] \\ \left(\downarrow R_{C_N} = f(R_{C_K}) \right) \\ C_N: f(S_{Pr} \times S_{Po} \times S_K \times S_H) \\ \left(\downarrow R_{C_U} = f(R_{C_N}) \right) \\ C_U: [P, V, M, C] \end{array} \right] \text{Light} \\[2em] \left[\begin{array}{c} M_3: -\infty \le t \le \infty \\ \downarrow \\ M_2: 0 \ge t > \infty \\ \downarrow \\ M_1: 0 > t > \infty \\ \downarrow \\ U \to \begin{array}{l} t \le E_{Cell}; TC: M_3 \to U \\ t \sim E_{Human}; TC: U \to M_3 \end{array} \end{array} \right] \text{Time} \end{array} \right. \right.
$$

$$
\left[\begin{array}{c} M_3 \to System_X \\ (\uparrow F \to I) \\ M_2 \to S_{System_X} \\ (\uparrow Sig \to F) \\ M_1 \to Sig_x \\ (\uparrow > P_x) \\ U \to x_{Established\ Matter} \end{array} \right] \text{Space}
$$

$$
TC \to x_{Super-Matter} \qquad \langle x_U | x_T \rangle
$$

Eq 4.1.11: Super-Matter Emergence

The emergence of super matter is discussed in greater detail in in the book Super-Matter (Malik, 2018a), with (4.1.11) included here to suggest computational realities yet to emerge.

SECTION 5: COMPUTING THE ELECTROMAGNETIC SPECTRUM

Section 5 will explore the computation involved in the creation of the electromagnetic spectrum. The electromagnetic spectrum is a technical way to refer to light, essentially because amongst its range of properties it displays a tangible and simultaneous electric and magnetic or "electromagnetic" reality as well. So as light becomes more concrete to us or as the possibilities within it begin to emerge, one of the first forms it takes is as the electromagnetic spectrum. It must be the case that the previously surfaced properties of light – Presence, Power, Knowledge, and Harmony – emerge so as to define the very architecture of the electromagnetic spectrum.

Chapter 5.1, Quantum Computations in Emergence of Electromagnetic Spectrum Logic, summarizes the emergence of the electromagnetic spectrum in terms of the underlying Light-Space-Time Emergence equation and the process of quantization that must occur to create the logic of the electromagnetic spectrum ecosystem that precipitates into the material-fabric. So we find that the four underlying properties of Light that we call Harmony, Knowledge, Power, and Presence are of the essence of the speed with which the electromagnetic spectrum moves, the wave-range within the electromagnetic spectrum, the energy-gradient within the electromagnetic spectrum, and the mass-possibilities due to the electromagnetic spectrum, respectively. Further the description – electromagnetic – seems to have captured the Power-Harmony aspects implicit in light. In reality the electro-magnetic spectrum can likely be more completely described as electro-magnetic-wavearchetype-masspotential spectrum.

Chapter 5.1: Quantum-Level Computation in Creation of an Electromagnetic Spectrum Ecosystem

As discussed previously the Light-Space-Time Emergence equation (3.1.3) being iterative, can be used to model emergence as it proceeds from simpler four-fold to more complex four-fold manifestations. But further, as implied by (3.6.5), the Quantization Effect of Organization on Material-Fabric equation, any process of organization has the possibility of altering the material-fabric or fabric of existence so long as the bases involved are driven primarily by a meta-level. The level of granularity in any process of quantum computation is proposed as being such a 'structure' or 'fabric' of organization that informs the material. Whilst such a fabric involves the material realm, it also involves layers antecedent to it, which must be included for superposition and entanglement to be considered correctly, and subsequently for any quantum-based computation to proceed more accurately.

Application of Light-Space-Time Emergence Equation

As suggested by (3.1.3), reproduced below for convenience, the architecture and details of the electromagnetic spectrum can be seen to be the result of the application of the Light, Space, and Time matrices as will be elaborated:

$$
Emergence_{light-space-time} =
\begin{Vmatrix}
\begin{matrix}
C_{\infty}:[Pr, Po, K, H] \\
\left(\downarrow R_{C_K} = f(R_{C_\infty}) \right) \\
C_K: [S_{Pr}, S_{Po}, S_K, S_H] \\
\left(\downarrow R_{C_N} = f(R_{C_K}) \right) \\
C_N: f\left(S_{Pr} \times S_{Po} \times S_K \times S_H\right) \\
\left(\downarrow R_{C_U} = f(R_{C_N}) \right) \\
C_U: [P, V, M, C]
\end{matrix}
\Bigg|_{Light}
\;
\begin{matrix}
M_3 \to System_X \\
(\uparrow F \to I) \\
M_2 \to S_{System_X} \\
(\uparrow Sig \to F) \\
M_1 \to Sig_x \\
(\uparrow \, > P_x) \\
U \to x_U
\end{matrix}
\Bigg|_{Space}
\\[2ex]
\begin{matrix}
M_3: -\infty \le t \le \infty \\
\downarrow \\
M_2: 0 \ge t > \infty \\
\downarrow \\
M_1: 0 > t > \infty \\
\downarrow \\
U \to \begin{matrix} t \le E_{Cell}; TC: M_3 \to U \\ t \sim E_{Human}; TC: U \to M_3 \end{matrix}
\end{matrix}
\Bigg|_{Time}
TC \to x_T
\end{Vmatrix}
\langle x_U \,|\, x_T \rangle
$$

Starting with the Light-Matrix, the top left-hand matrix in (3.1.3), the first line from the top, $C_\infty:[Pr, Po, K, H]$, specifies the fundamental architecture of the electromagnetic spectrum. As will be elaborated in the subsections on Harmony, Knowledge, Power, and Presence, each of these aspects are an emergence of the fundamental properties of Light at ∞.

Line 3 in the Light-Matrix, $C_K: [S_{Pr}, S_{Po}, S_K, S_H]$, elaborates the sets for Presence, Power, Knowledge, and Harmony, each containing multiple elements. For example, as will be explored in the subsection on Harmony, the following elements: 'connection', 'growing into one's own', 'form bonds', for example, are suggested to exist in the Set of Harmony or Nurturing, and are accessed to contribute to the operative reality so set up by virtue of light traveling at c.

Specifically, Line 5, $C_N: f(S_{Pr} \times S_{Po} \times S_K \times S_H)$, suggests that unique seeds are created from a combination of the elements from all four sets, with a particular element leading, that in effect creates the distinctness of the vast variety possible in the electromagnetic spectrum.

Line 6, $(\downarrow R_{C_U} = f(R_{C_N}))$, specifies quantization between the layer where the seeds are formed, and the physical layer we are familiar with, and as explored in Chapter 3.5 and 3.6, will result in Line 7, $C_U: [P, V, M, C]$, hence changing the material-fabric of existence. The possibilities represented by Lines 1 through 5 hence concretize through the quantization represented by Line 6 to become the electromagnetic spectrum with its physical (related to Presence), vital (related to Power), mental (related to Knowledge), and connection (related to Harmony) aspects now existing in material reality typified by Light moving at c. Note that just as Line 6 represents a process of quantization relating the layer of reality created by Light traveling at c with the antecedent layers, so too Lines 2 and 4 as previously discussed, also represent quantization of a more subtle kind that ultimately plays a critical part in allowing the material-fabric to express infinite diversity.

Typically it is the process as captured by the Space-Matrix that will determine if Line 6 is activated. Specifically patterns at the untransformed layer, U, will need to be overcome, as specified by the second-line from the bottom of the Space-Matrix: $(\uparrow > P_x)$. But as specified by the bottom-line of the Time-Matrix, reproduced below, it is only with the advent of the human-system that the automaticity of the action of meta-levels is reversed:

$$U \rightarrow \begin{array}{l} t \leq E_{Cell}; TC: M_3 \rightarrow U \\ t \sim E_{Human}; TC: U \rightarrow M_3 \end{array}$$

Hence in the case of the electromagnetic spectrum system, which in this emergence is a pre-human system, the fact that patterns do not need to be overcome means that quantization happens automatically.

Application of Quantization Effect of Organization on Material-Fabric Equation

Nonetheless, and given this context it is useful to review
Equation 3.6.5, Quantization Effect of Organization on Material-Fabric:

$$Impact\ on\ Materal - Fabric = \begin{bmatrix} |[L][S][T]TC \rightarrow x_T|_{(x_U \mid x_T)} \\ \times \\ (|mod(Z_Q)|_{Y > U}) \\ \ni \\ Z_Q, Z \in U\ (Space, Time, Energy, Gravity) \end{bmatrix}$$

Line 1 in the matrix is simply (3.6.1) the Simplified Light-Space-Time Emergence equation. To understand the quantization that may occur, the organization resulting from the application of Line 1 is multiplied ($\times$) by a modulation (mod) of a fourfold space-time-energy-gravity quantization (Z_Q), so long as Y>U, that is, the operative bases are relatively transformed and hence have an active influence greater than U. The fourfold quantization is specified by ($\ni Z_Q, Z \in U\ (Space, Time, Energy, Gravity)$) where ($\ni$) the quantization being applied (Z) is each (U) of the elements of the set (Space, Time, Energy, Gravity).

But as just summarized in the Time-Matrix in (3.1.3) Y is by definition greater than U and hence quantization is automatic. In terms of the electromagnetic spectrum such quantization implies that wholeness becomes fully active through specific space, time, energy, and gravity quantization to create an holistic "ecosystem" with its own "electromagnetic spectrum logic" as it were. The wholeness has now precipitated into the material-fabric and is available to be consciously and unconsciously tapped into.

Precipitation of "Magnetic" Aspect of Electromagnetic Spectrum into Material-Fabric

To begin with, we had already looked at how the speed of light sets up the nature of reality because of its speed. So for light traveling at c, past, present, future, and the notion of separation due to creation of islands of matter is the reality. It seems then that c architects the possibility of interaction in our system and can therefore be thought of a projection from the property of Harmony to create the basis for a matter-based harmony. So being, it can be suggested that the nature of the resultant interactions allows matter-based organizations, regardless of scale, to come into their own, to grow into their boundaries, and to form bonds based on the sense of being separated from other perceived organizations. This notion of forming bonds seems also to be related to the "magnetic" in electromagnetic.

In equation form, as in Equation 5.1.1 it would therefore be possible to specify the nature of reality so set up by the electromagnetic spectrum (EM_Spectrum) moving at c:

$$EM\_Spectrum_{Speed} = Xa + Y\bar{b}_{0-n} \ \ where \begin{bmatrix} X \in [S_{System_N}] \\ Y \in [S_{System_{Pr}}, S_{System_P}, S_{System_K}, S_{System_N}] \\ a, b \ are \ integers; a > b \end{bmatrix}$$

Eq 5.1.1: Speed of EM Spectrum

The notion of 'connection', 'growing into one's own', 'form bonds', amongst other attributes of such a reality can be seen as elements of the four sets of architectural forces. Hence 5.1.1 suggests the mathematical equation that so defines the nature of reality due to the speed of c of the electromagnetic spectrum.

Note that (5.1.1) already implies that Lines 1 – 5 in the Light Matrix (3.1.3) have been activated, and that the logic of the "magnetic" in the electro-magnetic-ecosystem will automatically precipitate into the material-fabric through the action of Line 6-7 of (3.1.3).

This also implies that as the speed of c changes, and as discussed in Chapter 2.1, the nature of relationship in such a reality will also change. Perhaps it is that there are several different relationships, or types of harmonies possible.

Precipitation of Wave Archetype Aspect of Electromagnetic Spectrum into Material-Fabric

The electromagnetic spectrum contains a range of waves embedded in it. These waves enable many different applications with practical utility apparent in every day life. So for example there is a region in the electromagnetic spectrum that we call radio waves, and others that we know as microwave, infrared visible light, ultraviolet, x-rays and gamma rays that each make possible many technologies that we use every single day. These ranges essentially code a range of technological-possibility, and therefore the wave-range within the electromagnetic spectrum can be thought of as encoding or of expressing some kind of precipitation or projection or emergence of the property of Knowledge. Or put another way, the property of Knowledge that is found in light, emerges as the wave-range implicit in the electromagnetic spectrum.

The EM spectrum itself, with its vast range of natures from gamma rays through visible light through radio waves, with its implicitness of time-space possibility as suggested by frequency (ν) and wavelength (λ), may be thought of as an arrangement of archetypes of what is possible in systems, and therefore is perhaps a precipitation of system-knowledge, as in Equation 5.1.2:

$$EM\_Spectrum_{Structure} = Xa + Yb_{0-n}^{-} \quad where \left[\begin{array}{c} X \in [S_{System_K}] \\ Y \in [S_{System_{Pr}}, S_{System_p}, S_{System_K}, S_{System_N}] \\ a, b \ are \ integers; a > b \end{array} \right]$$

Eq 5.1.2: Structure of EM Spectrum

Note that (5.1.2) already implies that Lines 1 – 5 in the Light Matrix (3.1.3) have been activated, and that the logic of the "wave-archetype" in the electro-magnetic-wavearchetype-masspotential ecosystem will automatically precipitate into the material-fabric through the action of Line 6-7 of (3.1.3).

What this also implies is that the significance or intent of the different types of waves that exist can also be expressed by this general equation where the X and Y elements will vary. What precisely these elements are will need to be worked out. A few representative equations, Equations 5.1.3 through 5.1.6 follow:

$$Gamma\ Rays_{Intent} = Xa + Y\bar{b}_{0-n} \quad where \begin{bmatrix} X \in [S_{System_K}] \\ Y \in [S_{System_{Pr}}, S_{System_P}, S_{System_K}, S_{System_N}] \\ a, b\ are\ integers; a > b \end{bmatrix}$$

Eq 5.1.3: Gamma Rays Intent

Knowing that some of the applications of Gamma Rays are in sterilizing and radiotherapy, these can be attributed as elements to the architectural sets. The totality of the use of can be thought of as the 'intent' of Gamma Rays. An understanding of the uses will allow a full set of the secondary elements to be mapped thus allowing (5.1.3) to be reverse-engineered.

Similarly each of the archetypes present in the electromagnetic spectrum can be mapped out. Sample mapping of intent include x-rays, infrared, and microwaves:

$$X - Rays_{Intent} = Xa + Y\bar{b}_{0-n} \quad where \begin{bmatrix} X \in [S_{System_K}] \\ Y \in [S_{System_{Pr}}, S_{System_P}, S_{System_K}, S_{System_N}] \\ a, b\ are\ integers; a > b \end{bmatrix}$$

Eq 5.1.4: X-Rays Intent

$$Infrared_{Intent} = Xa + Y\bar{b}_{0-n} \quad where \begin{bmatrix} X \in [S_{System_K}] \\ Y \in [S_{System_{Pr}}, S_{System_P}, S_{System_K}, S_{System_N}] \\ a, b\ are\ integers; a > b \end{bmatrix}$$

Eq 5.1.5: Infrared Intent

$$Microwaves_{Intent} = Xa + Y\bar{b}_{0-n} \quad where \begin{bmatrix} X \in [S_{System_K}] \\ Y \in [S_{System_{Pr}}, S_{System_P}, S_{System_K}, S_{System_N}] \\ a, b\ are\ integers; a > b \end{bmatrix}$$

Eq 5.1.6: Microwaves Intent

Precipitation of "Electro" Aspect of Electromagnetic Spectrum in Material-Fabric

The range of different wave-types or wavelengths implicit in the electromagnetic spectrum also moves with different frequencies. Since the speed of light is a constant, and it is known that speed is the product of frequency and wavelength, the greater the wavelength the lower will be the frequency of the wave-type. Conversely the less the wavelength the higher will be the frequency of the wave-type. And energy or power of a wave-type will depend on its frequency.

Hence, gamma rays that have a lower wavelength will have a higher frequency, and higher energy associated with it. Radio waves on the other hand that have a higher wavelength will have a lower frequency and therefore lower energy associated with it. So implicit in the electromagnetic spectrum is a gradient of energy. But it is also known that the penetration power is dependent on frequency. Therefore, the higher the frequency or energy, the higher will be the power of the wave-type. Another way to say this is that the property of Power in Light emerges as the energy-gradient in the electromagnetic spectrum. This Power aspect, by the way, seems to have been captured by the "electro" in electromagnetic.

The energy-gradient implicit in the EM spectrum suggests the power and energy with which knowledge moves and is perhaps a precipitation of system-power, as in Equation 5.1.7:

$$EM\_Spectrum_{Energy} = Xa + Y\bar{b}_{0-n} \quad where \begin{bmatrix} X \in [S_{System_p}] \\ Y \in [S_{System_{Pr}}, S_{System_p}, S_{System_K}, S_{System_N}] \\ a, b \ are \ integers; a > b \end{bmatrix}$$

Eq 5.1.7: EM Spectrum Energy

Note that (5.1.7) already implies that Lines 1 – 5 in the Light Matrix (3.1.3) have been activated, and that the logic of the "electro" in the electro-magnetic-wavearchetype-masspotential ecosystem will automatically precipitate into the material-fabric through the action of Line 6-7 of (3.1.3).

This energy as suggested is directly proportional to the frequency, of which there is an infinite range as predicted by Maxwell's equations. Different frequencies have different penetration profiles (HyperPhysics, 2016) and it may be suggested that the nature of the energy is also a precipitation of a range of to be determined meta-functions as suggested in some sample Equations 5.1.8 through 5.1.12. Hence:

Gamma Rays $_{Nature \ of \ Energy}$ $=$

$$Xa + Yb^-_{0-n} \quad where \left[\begin{array}{c} X \in [S_{System_P}] \\ Y \in [S_{System_{Pr}}, S_{System_P}, S_{System_K}, S_{System_N}] \\ a, b \ are \ integers; a > b \end{array} \right]$$

Eq 5.1.8: Gamma Rays Nature of Energy

$$X - Rays_{Nature \ of \ Energy} \ =$$

$$Xa + Yb^-_{0-n} \quad where \left[\begin{array}{c} X \in [S_{System_P}] \\ Y \in [S_{System_{Pr}}, S_{System_P}, S_{System_K}, S_{System_N}] \\ a, b \ are \ integers; a > b \end{array} \right]$$

Eq 5.1.9: X-Rays Nature of Energy

$$Ultraviolet_{Nature \ of \ Energy} \ =$$

$$Xa + Yb^-_{0-n} \quad where \left[\begin{array}{c} X \in [S_{System_P}] \\ Y \in [S_{System_{Pr}}, S_{System_P}, S_{System_K}, S_{System_N}] \\ a, b \ are \ integers; a > b \end{array} \right]$$

Eq 5.1.10: Ultraviolet Nature of Energy

$$White \ Light_{Nature \ of \ Energy} \ =$$

$$Xa + Yb^-_{0-n} \quad where \left[\begin{array}{c} X \in [S_{System_P}] \\ Y \in [S_{System_{Pr}}, S_{System_P}, S_{System_K}, S_{System_N}] \\ a, b \ are \ integers; a > b \end{array} \right]$$

Eq 5.1.11: White Light Nature of Energy

$$AM \ Radio \ Waves_{Nature \ of \ Energy} \ =$$

$$Xa + Yb^-_{0-n} \quad where \left[\begin{array}{c} X \in [S_{System_P}] \\ Y \in [S_{System_{Pr}}, S_{System_P}, S_{System_K}, S_{System_N}] \\ a, b \ are \ integers; a > b \end{array} \right]$$

Eq 5.1.12: AM Radio Waves Nature of Energy

Precipitation of Mass Potential Aspect of Electromagnetic Spectrum into Material-Fabric

If there is a large range of frequencies implicit in the electromagnetic spectrum then there is also the possibility of different types of masses implicit in the electromagnetic spectrum. Frequency determines energy, and mass and energy are related through Einstein's famous MC-squared equation. So pushing a little further it is not just that mass and energy are related, but a different kind of frequency or wave-type potentially allows a different type of mass to emerge. So the possibility of different types of mass seems to be related to the property of Presence in Light. In other words the property of Presence emerges as the possibility of different types of masses as suggested by the range of mass-possibilities that can emerge from the electromagnetic spectrum.

Mass can be thought of as a container at U within which all possibility happens. In other words it can be thought of as a precipitation or emergence of system-presence as depicted in Equation 5.1.13:

$$EM_{SpectrumMass_{Possibility}} =$$

$$Xa + Yb^-_{0-n} \ \ where \begin{bmatrix} X \in [S_{System_{Pr}}] \\ Y \in [S_{System_{Pr}}, S_{System_{P}}, S_{System_{K}}, S_{System_{N}}] \\ a, b \ are \ integers; a > b \end{bmatrix}$$

Eq 5.1.13: EM Spectrum Mass Possibility

Note that (5.1.13) already implies that Lines 1 – 5 in the Light Matrix (3.1.3) have been activated, and that the logic of the "masspotential" in the electro-magnetic-wavearchetype-masspotential ecosystem will automatically precipitate into the material-fabric through the action of Line 6-7 of (3.1.3).

Further, if the frequencies are infinite, then the possibility of the 'types' of masses or matter is also infinite. The mystery of 'Dark Matter' suggested by scientists to be 27% of our universe, as opposed to 5% of visible matter (NASA-darkmatter, 2016) may have some relation to this. This aspect is also explored in the book Cosmology of Light (Malik, 2018b).

Hypothetically the Equations 5.1.14 through 5.1.18 depict a range of mass possibilities:

$$Gamma\ Rays\ _{Mass_{Possibility}} =$$

$$Xa + Yb_{0-n}^{-}\ where \begin{bmatrix} X \in [S_{System_{Pr}}] \\ Y \in [S_{System_{Pr}}, S_{System_P}, S_{System_K}, S_{System_N}] \\ a, b\ are\ integers; a > b \end{bmatrix}$$

Eq 5.1.14: Gamma Rays Mass Possibility

$$Ultraviolet\ _{Mass_{Possibility}} =$$

$$Xa + Yb_{0-n}^{-}\ where \begin{bmatrix} X \in [S_{System_{Pr}}] \\ Y \in [S_{System_{Pr}}, S_{System_P}, S_{System_K}, S_{System_N}] \\ a, b\ are\ integers; a > b \end{bmatrix}$$

Eq 5.1.15: Ultraviolet Mass Possibility

$$Blue\ Light\ _{Mass_{Possibility}} =$$

$$Xa + Yb_{0-n}^{-}\ where \begin{bmatrix} X \in [S_{System_{Pr}}] \\ Y \in [S_{System_{Pr}}, S_{System_P}, S_{System_K}, S_{System_N}] \\ a, b\ are\ integers; a > b \end{bmatrix}$$

Eq 5.1.16: Blue Light Mass Possibility

$$Microwaves\ _{Mass_{Possibility}} =$$

$$Xa + Yb_{0-n}^{-}\ where \begin{bmatrix} X \in [S_{System_{Pr}}] \\ Y \in [S_{System_{Pr}}, S_{System_P}, S_{System_K}, S_{System_N}] \\ a, b\ are\ integers; a > b \end{bmatrix}$$

Eq 5.1.17: Microwaves Mass Possibility

$$FM\ Radio\ Waves_{Mass_{Possibility}} =$$

$$Xa + Yb_{0-n}^{-}\ where \begin{bmatrix} X \in [S_{System_{Pr}}] \\ Y \in [S_{System_{Pr}}, S_{System_P}, S_{System_K}, S_{System_N}] \\ a, b\ are\ integers; a > b \end{bmatrix}$$

Eq 5.1.18: FM Radio Waves Mass Possibility

SECTION 6: COMPUTING MATTER

So we find that the four underlying properties of Light that we call Harmony, Knowledge, Power, and Presence are of the essence of the speed with which the electromagnetic spectrum moves, the wave-range within the electromagnetic spectrum, the energy-gradient within the electromagnetic spectrum, and the mass-possibilities due to the electromagnetic spectrum, respectively. Further the description – electromagnetic – seems to have captured the Power-Harmony aspects implicit in light. In reality the electro-magnetic spectrum can likely be more completely described as electro-magnetic-wavearchetype-masspotential spectrum.

But further, we will find that layers of matter – quantum particles, which include bosons, and atoms – are also structured or emerge along the same property-lines or property-families of light.

There is similarly a continuous process of computation that involves the quantum-realms and quantization to create the realities of quantum particles, including bosons, and atoms.

Hence the following chapters, 6.1, 6.2, and 6.3, describe a process of computation by which Light emerges as quantum particles, bosons, and atoms respectively.

The electromagnetic spectrum or perhaps more accurately, the electro-magnetic-wavearchetype-masspotential spectrum, describes one of the first layers or translation of light into manifestation. This spectrum or field is an example of what stands behind visible matter and perhaps it is fair to say that all matter emerges from such spectrums.

As already explored the finite speed of such spectrums allows the build-up of energies or quanta, and this in turn may express itself as a series or as an array of quantum particles. The vast number of quantum particles that have so far been discovered has in turn yielded what is known as the Standard Model. This model is made up of what is called Quarks, Leptons, and Bosons, and a Higgs-Boson (Cottingham, 2007).

But if we look at these four fundamental quantum categories of particles from a property-based, or "functional" viewpoint a different chapter in the exploration of light emerges. All of matter then can be seen as implicit in Light and in fact due to a process of computation by which the implicit codification in Light becomes explicit. The process of becoming explicit involves multiple layers of light, superposition, entanglement, and quanta. There is a major on-going computation that is of the very nature of existence, as elaborated in Section 2, that draws out more possibilities in Light through a process of dynamic interrelation between what has already become material and what is held in potential in Light's codification, as elaborated in Sections 3, 4, and 5.

The on-going computation also involves a process of quantization by which the very logic of quantum particles precipitates into the material-fabric and the material layer specified by U.

Light's Emergence as Quantum Particles

As discussed previously the Light-Space-Time Emergence equation (3.1.3) being iterative, can be used to model emergence as it proceeds from simpler four-fold to more complex four-fold manifestations. But further, as implied by (3.6.5), the Quantization Effect of Organization on Material-Fabric equation, any process of organization has the possibility of altering the material-fabric or fabric of existence so long as the bases involved are driven primarily by a meta-level.

128

As suggested by (3.1.3), reproduced below for convenience, the architecture and details of quantum particles can be seen to be the result of the application of the Light, Space, and Time matrices as will be elaborated:

$Emergence_{light - space - time} =$

$$
\left|
\begin{array}{l}
\left|
\begin{array}{c}
C_\infty: [Pr, Po, K, H] \\
\left(\downarrow R_{C_K} = f(R_{C_\infty}) \right) \\
C_K: [S_{Pr}, S_{Po}, S_K, S_H] \\
\left(\downarrow R_{C_N} = f(R_{C_K}) \right) \\
C_N: f(S_{Pr} \times S_{Po} \times S_K \times S_H) \\
\left(\downarrow R_{C_U} = f(R_{C_N}) \right) \\
C_U: [P, V, M, C]
\end{array}
\right|_{Light}
\left|
\begin{array}{c}
M_3 \to System_X \\
(\uparrow F \to I) \\
M_2 \to S_{System_X} \\
(\uparrow Sig \to F) \\
M_1 \to Sig_X \\
(\uparrow > P_x) \\
U \to x_U
\end{array}
\right|_{Space} \\
\left|
\begin{array}{c}
M_3: -\infty \le t \le \infty \\
\downarrow \\
M_2: 0 \ge t > \infty \\
\downarrow \\
M_1: 0 > t > \infty \\
\downarrow \\
U \to \begin{array}{l} t \le E_{Cell}; TC: M_3 \to U \\ t \sim E_{Human}; TC: U \to M_3 \end{array}
\end{array}
\right|_{Time} TC \to x_T
\end{array}
\right| \langle x_U | x_T \rangle
$$

Starting with the Light-Matrix, the top left-hand matrix in (3.1.3), the first line from the top, $C_\infty: [Pr, Po, K, H]$, specifies the fundamental architecture of quantum particles. In this rendering, and as elaborated in subsequent subsections, quarks are an emergence of Light's property of Knowledge, leptons are an emergence of Light's property of Power, bosons are an emergence of Light's property of Harmony, and the Higgs-boson is an emergence of Light's property of Presence. The fundamental architecture of these aspects hence, is an emergence of the properties of Light at ∞.

Line 3 in the Light-Matrix, $C_K: [S_{Pr}, S_{Po}, S_K, S_H]$, elaborates the sets for Presence, Power, Knowledge, and Harmony, each containing multiple elements. For example, as will be explored in greater detail in the section on quarks, various elements derived from the four sets define the behavior of quarks and could be functions such as 'composite-arrangements', 'specifying attributes' amongst others, hence collectively describing quarks' way of being. Specifically, Line 5, $C_N: f(S_{Pr} \times S_{Po} \times S_K \times S_H)$, suggests that unique seeds are created from a combination of such elements from all four sets, with a particular element

129

leading, that in effect creates the distinctness possible at the level of quantum particles.

Line 6, $(\downarrow R_{C_U} = f(R_{C_N}))$, specifies quantization between the layer where the seeds are formed, and the physical layer we are familiar with, and as explored in Chapter 3.5 and 3.6, will result in Line 7, C_U: $[P,V,M,C]$, hence changing the material-fabric of existence. The possibilities represented by Lines 1 through 5 hence concretize through the quantization represented by Line 6 to become the quantum particles with its physical (related to Presence), vital (related to Power), mental (related to Knowledge), and connection (related to Harmony) aspects now existing in material reality typified by Light moving at c. Note that just as Line 6 represents a process of quantization relating the layer of reality created by Light traveling at c with the antecedent layers, so too Lines 2 and 4 as previously discussed, also represent quantization of a more subtle kind that ultimately plays a critical part in allowing the material-fabric to express infinite diversity.

Typically it is the process as captured by the Space-Matrix that will determine if Line 6 is activated. Specifically patterns at the untransformed layer, U, will need to be overcome, as specified by the second-line from the bottom of the Space-Matrix: $(\uparrow > P_x)$. But as specified by the bottom-line of the Time-Matrix, reproduced below, it is only with the advent of the human-system that the automaticity of the action of meta-levels is reversed:

$$U \rightarrow \begin{array}{l} t \leq E_{Cell}; TC: M_3 \rightarrow U \\ t \sim E_{Human}; TC: U \rightarrow M_3 \end{array}$$

Hence in the case of the quantum particle system, which in this emergence is a pre-human system, the fact that patterns do not need to be overcome means that quantization happens automatically.

Nonetheless, and given this context it is useful to review
Equation 3.6.5, Quantization Effect of Organization on Material-Fabric:

$$Impact\ on\ Materal-Fabric = \begin{bmatrix} |[L][S][T]TC \rightarrow x_T|_{\langle x_U | x_T \rangle} \\ \times \\ (|mod(Z_Q)|_{Y>U}) \\ \ni \\ Z_Q, Z \in U\ (Space, Time, Energy, Gravity) \end{bmatrix}$$

Line 1 in the matrix is simply (3.6.1) the Simplified Light-Space-Time Emergence equation. To understand the quantization that may occur, the organization resulting from the application of Line 1 is multiplied $(\times)$ by a modulation (mod) of a fourfold space-time-energy-gravity quantization (Z_Q), so long as Y>U, that is, the operative bases are relatively transformed and hence have an active influence greater than U. The fourfold quantization is specified by $(\ni Z_Q, Z \in U \ (Space, Time, Energy, Gravity))$ where ($\ni$) the quantization being applied (Z) is each (U) of the elements of the set (Space, Time, Energy, Gravity).

But as just summarized in the Time-Matrix in (3.1.3) Y is by definition greater than U and hence quantization is automatic. In terms of quantum particles such quantization implies that wholeness becomes fully active through specific space, time, energy, and gravity quantization to create an holistic "ecosystem" with its own "quantum particle logic" as it were. The wholeness has now precipitated into the material-fabric and is available to be consciously and unconsciously tapped into.

Precipitation of Quark Logic into Material-Fabric

It can be seen first that the nucleus of an atom is made up of a combination of quarks. Specifically, a proton is composed of two "up" quarks and one "down" quark. Quarks have unusual names – up, down, charm, strange, top, bottom, with each subsequent pair belonging to a different "generation" of quarks. A neutron is composed of two "down" quarks and one "up" quark. Protons and neutrons together make up the nucleus of an atom. But we also know that the number of protons in the nucleus specifies the Atomic Number of an atom. Atomic number in turn uniquely identifies the element from the periodic table. Hence, an atomic number of 47, for example, specifies that the element is Silver. In other words it can be suggested that the unique properties of an element, the knowledge of what it is and how it will behave in the universe, is related to the quark. It may be suggested that quarks, therefore, are associated with the precipitation or emergence of Light's property of Knowledge in the quantum world.

Hence, it could be that the signature for the family of quarks, as in Equation 6.1.1, is:

$$Sig_{quarks} = Xa + Yb_{0-n} \quad where \begin{bmatrix} X \in [S_{System_K}] \\ Y \in [S_{System_{Pr}}, S_{System_p}, S_{System_K}, S_{System_N}] \\ a, b \ are \ integers; a > b \end{bmatrix}$$

Eq 6.1.1: Generalized Signature of Quarks

As can be seen the primary element X is derived from the set of knowledge, S_{System_K}. Various elements, derived from the four sets would define the behavior of quarks and could be functions such as 'composite-arrangements', 'specifying attributes' amongst others, hence collectively describing quarks' way of being.

Note that (6.1.1) already implies that Lines 1 – 5 in the Light Matrix (3.1.3) have been activated, and that the logic of the quark-ecosystem will automatically precipitate into the material-fabric through the action of Line 6-7 of (3.1.3).

Precipitation of Lepton Logic into Material-Fabric

When considering leptons it is useful to know that unlike quarks that only exist in composite arrangements with other quarks, leptons are solitary, point-like particles without internal structure (Arabatzis, 2006). The best-known lepton is the electron. So the electron may be considered as a surrogate for the lepton class. The electron appears to be associated with the flow of energy and power. Further they appear to be the adventurers easily leaving the atom they are a part of. They also form locks or bonds with other atoms through the force of attraction and repulsion. In some sense they seem to be a representation or precipitation or emergence of Light's property of Power.

The signature for the family of leptons, as in Equation 6.1.2, is:

$$Sig_{leptons} = Xa + Yb_{0-n} \quad where \begin{bmatrix} X \in [S_{System_p}] \\ Y \in [S_{System_{Pr}}, S_{System_p}, S_{System_K}, S_{System_N}] \\ a, b \ are \ integers; a > b \end{bmatrix}$$

Eq 6.1.2: Generalized Signature of Leptons

As can be seen the primary element X is derived from the set of power, S_{System_P}. Various elements, derived from the four sets would define the behavior of leptons and could be functions such as 'adventurer', 'solitary',

132

'create attraction', amongst others, hence collectively describing leptons' way of being.

Note that (6.1.2) already implies that Lines 1 – 5 in the Light Matrix (3.1.3) have been activated, and that the logic of the lepton-ecosystem will automatically precipitate into the material-fabric through the action of Line 6-7 of (3.1.3).

Precipitation of Boson Logic into Material-Fabric

The bosons on the other hand are thought of as force-carriers. They are what allow all known matter particles to interact. The three fundamental bosons in this category are the photon, the W and Z bosons, and the gluon. The carrier particle of the electromagnetic force or spectrum is the photon. The carrier particle of the strong nuclear force that holds quarks together is the gluon. The carrier particle for the weak interactions, responsible for the decay of massive quarks and leptons into lighter quarks and leptons, are the W and Z bosons.

Bosons can be thought of as the precipitation of what creates relationship and harmony at the quantum level. Hence they can be thought of as the precipitation or emergence of Light's property of Harmony at the quantum level.

The signature for the family of gauge bosons, as in Equation 6.1.3, is:

$$Sig_{bosons} = Xa + Yb_{0-n}^- \quad where \begin{bmatrix} X \in [S_{System_N}] \\ Y \in [S_{System_{Pr}}, S_{System_p}, S_{System_K}, S_{System_N}] \\ a, b \ are \ integers; a > b \end{bmatrix}$$

Eq 6.1.3: Generalized Signature of Bosons

As can be seen the primary element X is derived from the set of nurturing, S_{System_N}. Various elements, derived from the four sets would define the behavior of bosons and could be functions such as 'creating interaction', 'holding together', amongst others, hence collectively describing bosons' way of being.

Note that (6.1.3) already implies that Lines 1 – 5 in the Light Matrix (3.1.3) have been activated, and that the logic of the boson-ecosystem will automatically precipitate into the material-fabric through the action of Line 6-7 of (3.1.3).

Equations in this form give a clue as to how to think about the fundamental particles. While the default approach is always to think about physical qualities, equations in the preceding format allow us to think about detailed functions or qualities associated with particles and give insight into how the process of emergence or complexification of fourfold functionality takes place.

Precipitation of Higgs-Boson Logic into Material-Fabric

This leaves the other discovered fundamental particle the Higgs-Boson. In ordinary matter, most of the mass is contained in atoms, and the majority of the mass of an atom resides in the nucleus, made of protons and neutrons. Protons and neutrons are each made of three quarks. But it is the quarks that get their mass by interacting with what is known as the Higgs field (Olive, 2014). Hence the Higgs-Boson can be thought of as the mass-giver. In other words it is what gives presence to the quarks and it can be thought of as the precipitation or emergence of Light's property of Presence at the quantum level. Just as there are multiple particles in each of the other 'families' it is likely that there will be multiple particles in the Higgs-Boson family. Recent research at CERN indicates that the Higgs-Boson may have a cousin.

The signature for the Higgs-boson and any other similar particle, as in Equation 6.1.4, is:

$$Sig_{Higgs-boson} = Xa + Y\bar{b}_{0-n} \quad where \begin{bmatrix} X \in [S_{System_{Pr}}] \\ Y \in [S_{System_{Pr}}, S_{System_p}, S_{System_K}, S_{System_N}] \\ a, b \ are \ integers; a > b \end{bmatrix}$$

Eq 6.1.4: Generalized Signature of Higgs-Boson

As can be seen the primary element X is derived from the set of presence, $S_{System_{Pr}}$. Various elements, derived from the four sets would define the behavior of Higgs-bosons and could be a function such as 'creating mass', amongst others, hence collectively describing Higgs-bosons' way of being.

Note that (6.1.4) already implies that Lines 1 – 5 in the Light Matrix (3.1.3) have been activated, and that the logic of the Higgs-boson-ecosystem will automatically precipitate into the material-fabric through the action of Line 6-7 of (3.1.3).

Combining all the preceding particle equations it is possible to create a generalized particle equation, as in Equation 6.1.5:

$$Sig_{particle} = Xa + Y\overline{b}_{0-n} \quad where \begin{bmatrix} X \in [S_{System_{Pr}}, S_{System_p}, S_{System_K}, S_{System_N}] \\ Y \in [S_{System_{Pr}}, S_{System_p}, S_{System_K}, S_{System_N}] \\ a, b \ are \ integers; a > b \end{bmatrix}$$

Eq 6.1.5: Generalized Signature of Particle

So we see that again the underlying properties of light - Knowledge, Power, Harmony, and Presence - emerge as quarks, leptons, bosons, and Higgs-bosons respectively.

Chapter 6.2: Quantum-Level Computation in Creation of a Boson Ecosystem

The bosons as mentioned can be thought of as force-carriers and allow all known matter particles to interact.

But when we look at bosons in more detail there are three fundamental bosons – the photon, the W and Z bosons, the gluon - and one hypothetical boson – the graviton.

This chapter looks at the on-going computation involving a process of quantization by which the very logic of bosons precipitates into the material-fabric specified by U.

Light's Emergence as Bosons

As discussed previously the Light-Space-Time Emergence equation (3.1.3) being iterative, can be used to model emergence as it proceeds from simpler four-fold to more complex four-fold manifestations. Hence (3.1.3) has already been applied to suggest the emergence of the electromagnetic spectrum and quantum particles in general. Here it will be applied to suggest the emergence of a sub-class of quantum particles, Bosons. But further, as implied by (3.6.5), the Quantization Effect of Organization on Material-Fabric equation, any process of organization, such as the architecture and cohesiveness of bosons when viewed with the lens provided here, has the possibility of altering the material-fabric or fabric of existence so long as the bases involved are driven primarily by a meta-level.

As suggested by (3.1.3), reproduced below for convenience, the architecture and details of bosons can be seen to be the result of the application of the Light, Space, and Time matrices as will be elaborated:

$$Emergence_{light-space-time} =$$

$$
\begin{array}{|c|c|c|}
\hline
\begin{matrix}
C_\infty : [Pr, Po, K, H] \\
\left(\downarrow R_{C_K} = f(R_{C_\infty})\right) \\
C_K : [S_{Pr}, S_{Po}, S_K, S_H] \\
\left(\downarrow R_{C_N} = f(R_{C_K})\right) \\
C_N : f\left(S_{Pr} \times S_{Po} \times S_K \times S_H\right) \\
\left(\downarrow R_{C_U} = f(R_{C_N})\right) \\
C_U : [P, V, M, C]
\end{matrix}
&
\begin{matrix}
M_3 \to System_X \\
(\uparrow F \to I) \\
M_2 \to S_{System_X} \\
(\uparrow Sig \to F) \\
M_1 \to Sig_x \\
(\uparrow > P_x) \\
U \to x_U
\end{matrix}
& \\
Light & Space & \\
\hline
\begin{matrix}
M_3 : -\infty \leq t \leq \infty \\
\downarrow \\
M_2 : 0 \geq t > \infty \\
\downarrow \\
M_1 : 0 > t > \infty \\
\downarrow \\
U \to \begin{array}{l} t \leq E_{Cell}; TC: M_3 \to U \\ t \sim E_{Human}; TC: U \to M_3 \end{array}
\end{matrix}
& Time \quad TC \to x_T
& \langle x_U \mid x_T \rangle \\
\hline
\end{array}
$$

Starting with the Light-Matrix, the top left-hand matrix in (3.1.3), the first line from the top, $C_\infty : [Pr, Po, K, H]$, specifies the fundamental architecture of bosons. As explored previously in this chapter, Gluons are an emergence of Light's property of Knowledge, W and Z bosons are an emergence of Light's property of Power, Photons are an emergence of Light's property of Presence, and the Graviton is an emergence of Light's property of Harmony. The fundamental architecture of these aspects hence, is an emergence of the properties of Light at ∞.

Line 3 in the Light-Matrix, $C_K : [S_{Pr}, S_{Po}, S_K, S_H]$, elaborates the sets for Presence, Power, Knowledge, and Harmony, each containing multiple elements. For example, as will be explored in the section on photons, various elements derived from the four sets define the behavior of photons and could be functions such as 'pervasiveness', 'multiple wave handler', amongst others, hence collectively describing photons' way of being. Specifically, Line 5, $C_N : f\left(S_{Pr} \times S_{Po} \times S_K \times S_H\right)$, suggests that unique seeds are created from a combination of such elements from all four sets, with a particular element leading, that in effect creates the distinctness possible at the level of bosons.

Line 6, $\left(\downarrow R_{C_U} = f(R_{C_N})\right)$, specifies quantization between the layer where the seeds are formed, and the physical layer we are familiar with, and as explored in Chapter 3.5 and 3.6, will result in Line 7, $C_U : [P, V, M, C]$, hence changing the material-fabric of existence. The possibilities represented by Lines 1 through 5 hence concretize through the quantization represented by Line 6 to become the bosons with its physical (related to Presence), vital (related to Power), mental (related to Knowledge), and connection (related to Harmony) aspects now existing in material reality typified by Light moving at c. Note that just as Line 6 represents a process of quantization relating the layer of reality created by Light traveling at c with the antecedent layers, so too Lines 2 and 4 as previously discussed, also represent quantization of a more subtle kind that ultimately plays a critical part in allowing the material-fabric to express infinite diversity.

Typically it is the process as captured by the Space-Matrix that will determine if Line 6 is activated. Specifically patterns at the untransformed layer, U, will need to be overcome, as specified by the second-line from the bottom of the Space-Matrix: $(\uparrow > P_x)$. But as specified by the bottom-line of the Time-Matrix, reproduced below, it is only with the advent of the human-system that the automaticity of the action of meta-levels is reversed:

$$U \rightarrow \begin{array}{l} t \le E_{Cell}; TC: M_3 \rightarrow U \\ t \sim E_{Human}; TC: U \rightarrow M_3 \end{array}$$

Hence in the case of bosons, which in this emergence is a pre-human system, the fact that patterns do not need to be overcome means that quantization happens automatically.

Nonetheless, and given this context it is useful to review Equation 3.6.5, Quantization Effect of Organization on Material-Fabric:

$$Impact\ on\ Materal\ Fabric = \begin{bmatrix} |[L][S][T]TC \rightarrow x_T|_{(x_U | x_T)} \\ \times \\ (|mod(Z_Q)|_{Y > U}) \\ \ni \\ Z_Q, Z \in U\ (Space, Time, Energy, Gravity) \end{bmatrix}$$

Line 1 in the matrix is simply (3.6.1) the Simplified Light-Space-Time Emergence equation. To understand the quantization that may occur, the organization resulting from the application of Line 1 is multiplied $(\times)$ by a modulation (mod) of a fourfold space-time-energy-gravity quantization (Z_Q), so long as Y>U, that is, the operative bases are relatively transformed and hence have an active influence greater than U. The fourfold quantization is specified by $(\ni Z_Q, Z \in U\ (Space, Time, Energy, Gravity))$ where $(\ni)$ the quantization

138

being applied (Z) is each $(^U)$ of the elements of the set (Space, Time, Energy, Gravity).

But as just summarized in the Time-Matrix in (3.1.3) Y is by definition greater than U and hence quantization is automatic. In terms of bosons such quantization implies that wholeness becomes fully active through specific space, time, energy, and gravity quantization to create an holistic "ecosystem" with its own "boson logic" as it were. The wholeness has now precipitated into the material-fabric and is available to be consciously and unconsciously tapped into.

Precipitation of Photon Logic into Material-Fabric

The photon is the carrier particle of the electromagnetic force. But in the scheme of things the electromagnetic force pervades everything and as explored in the section on the electromagnetic spectrum appears to be foundational to this reality we are in. So it could be said that it is related to Presence or is an emergence of Light's property of Presence at the level of quantum force-carriers.

Hence, an equation for the photon, Equation 6.2.1 could be:

$$Sig_{photon} = Xa + Yb^-_{0-n} \;\; where \begin{bmatrix} X \in [S_{System_{Pr}}] \\ Y \in [S_{System_{Pr}}, S_{System_P}, S_{System_K}, S_{System_N}] \\ a, b \; are \; integers; a > b \end{bmatrix}$$

Eq 6.2.1: Signature of Photon

While the primary element X, would continue to be the same as X for bosons, as in Equation 6.1.3, there will be additional secondary elements Y that will further qualify the attributes or functionality of photons. As an example such elements could be 'pervasiveness', 'multiple wave handler', amongst others.

Note that (6.2.1) already implies that Lines 1 – 5 in the Light Matrix (3.1.3) have been activated, and that the logic of the photon-ecosystem will automatically precipitate into the material-fabric through the action of Line 6-7 of (3.1.3).

Precipitation of Gluon Logic into Material-Fabric

The gluon is the carrier particle of what is known as the strong nuclear force and holds quarks together in their inherently composite arrangements. But we have posited that quarks are related to or are an emergence of Knowledge at the quantum-particle level. Hence it must be that the gluon is an emergence of Light's property of Knowledge at the level of quantum force-carriers.

Hence, Equation 6.2.2 is an equation for the gluon:

$$Sig_{gluon} = Xa + Yb_{0-n}^{-} \quad where \begin{bmatrix} X \in [S_{System_K}] \\ Y \in [S_{System_{Pr}}, S_{System_P}, S_{System_K}, S_{System_N}] \\ a, b \ are \ integers; a > b \end{bmatrix}$$

Eq 6.2.2: Signature of Gluon

The primary element X would continue to be the same as in the equation for the boson. There are likely additional secondary elements Y such as 'concentrated', 'intense connection', amongst others that would collectively determine the behavior of gluons.

Note that (6.2.2) already implies that Lines 1 – 5 in the Light Matrix (3.1.3) have been activated, and that the logic of the gluon-ecosystem will automatically precipitate into the material-fabric through the action of Line 6-7 of (3.1.3).

Precipitation of W and Z Bosons Logic into Material-Fabric

The W and Z bosons are the carrier particle for the weak interactions, responsible for the decay of massive quarks and leptons into lighter quarks and leptons. But this usually is accompanied by the release of energy and power and so it must be that W and Z bosons are an emergence of Light's property of Power at the level of quantum force-carriers.

Hence, Equation 6.2.3, for W and Z bosons is as follows:

$$Sig_{W,Z \ bosons} = Xa + Yb_{0-n}^{-} \quad where \begin{bmatrix} X \in [S_{System_P}] \\ Y \in [S_{System_{Pr}}, S_{System_P}, S_{System_K}, S_{System_N}] \\ a, b \ are \ integers; a > b \end{bmatrix}$$

Eq 6.2.3: Signature of W and Z Bosons

The primary element X would continue to be the same as in the equation for the boson. There are likely additional secondary elements Y such as 'weak link', 'release of energy', amongst others that would collectively determine the behavior of W and Z bosons.

140

Note that (6.2.3) already implies that Lines 1 – 5 in the Light Matrix (3.1.3) have been activated, and that the logic of the W and Z-bosons-ecosystem will automatically precipitate into the material-fabric through the action of Line 6-7 of (3.1.3).

Precipitation of Graviton Logic into Material-Fabric

This leaves the hypothetical graviton that is thought to be the carrier particle for the force of gravity. But gravity is what holds astronomical objects together in relationship. Hence it must be that gravitons are an emergence of Light's property of Harmony at the level of quantum force-carriers.

Hence, Equation 6.2.4, for the hypothetical graviton is as follows:

$$Sig_{graviton} = Xa + Yb^{-}_{0-n} \quad where \left[\begin{array}{c} X \in [S_{System_N}] \\ Y \in [S_{System_{Pr}}, S_{System_P}, S_{System_K}, S_{System_N}] \\ a, b \ are \ integers; a > b \end{array} \right]$$

Eq 6.2.4: Signature of Graviton

The primary element X would continue to be the same as in the equation for the boson. There are likely additional secondary elements Y such as 'mass curving space', 'space defining object movement', amongst others that would collectively determine the behavior of gravitons.

Note that (6.2.4) already implies that Lines 1 – 5 in the Light Matrix (3.1.3) have been activated, and that the logic of the graviton-ecosystem will automatically precipitate into the material-fabric through the action of Line 6-7 of (3.1.3).

So we see that the underlying properties of light - Presence, Knowledge, Power, and Harmony - emerge as photons, gluons, W and Z bosons, and the hypothetical graviton at the quantum force-carrier level.

Chapter 6.3: Quantum-Level Computation in Creation of an Atom Ecosystem

Quantum particles in turn create atoms, and all known atoms can be classified into one of four groups: The s-Group, p-Group, d-Group, and f-Group. But what are these groups and are they too emergences of Light as it continues its journey of crystalizing the possibility or potentiality or fourfold functionality within it?

Light's Emergence as Atoms

As discussed previously the Light-Space-Time Emergence equation (3.1.3) being iterative, can be used to model emergence as it proceeds from simpler four-fold to more complex four-fold manifestations. Hence (3.1.3) has already been applied to suggest the emergence of the electromagnetic spectrum, quantum particles in general, and bosons as a further instance of a particular kind of quantum particle. Here it will be applied to suggest the emergence of Atoms. But further, as implied by (3.6.5), the Quantization Effect of Organization on Material-Fabric equation, any process of organization, such as the architecture and cohesiveness of atoms, has the possibility of altering the material-fabric or fabric of existence so long as the bases involved are driven primarily by a meta-level.

As suggested by (3.1.3), reproduced below for convenience, the architecture and details of atoms can be seen to be the result of the application of the Light, Space, and Time matrices as will be elaborated:

$$Emergence_{light-space-time} =$$

$$\left|\begin{array}{l} C_{\infty}:[Pr, Po, K, H] \\ \left(\downarrow R_{C_K} = f(R_{C_\infty})\right) \\ C_K: [S_{Pr}, S_{Po}, S_K, S_H] \\ \left(\downarrow R_{C_N} = f(R_{C_K})\right) \\ C_N: f(S_{Pr} \times S_{Po} \times S_K \times S_H) \\ \left(\downarrow R_{C_U} = f(R_{C_N})\right) \\ C_U: [P, V, M, C] \end{array}\right|_{Light} \quad \left|\begin{array}{l} M_3 \to System_X \\ (\uparrow F \to I) \\ M_2 \to S_{System_X} \\ (\uparrow Sig \to F) \\ M_1 \to Sig_x \\ (\uparrow > P_x) \\ U \to x_U \end{array}\right|_{Space}$$

$$\left|\begin{array}{c} M_3: -\infty \le t \le \infty \\ \downarrow \\ M_2: 0 \ge t > \infty \\ \downarrow \\ M_1: 0 > t > \infty \\ \downarrow \\ U \to \begin{array}{l} t \le E_{Cell}; TC: M_3 \to U \\ t \sim E_{Human}; TC: U \to M_3 \end{array} \end{array}\right|_{Time} \quad TC \to x_T \qquad \langle x_U \,|\, x_T \rangle$$

Starting with the Light-Matrix, the top left-hand matrix in (3.1.3), the first line from the top, $C_{\infty}:[Pr, Po, K, H]$, specifies the fundamental architecture of atoms. As will be explored in this chapter, p-Group atoms are an emergence of Light's property of Knowledge, s-Group atoms are an emergence of Light's property of Power, d-Group atoms are an emergence of Light's property of Presence, and the f-Group atoms are an emergence of Light's property of Harmony. The fundamental architecture of these aspects hence, is an emergence of the properties of Light at ∞.

Line 3 in the Light-Matrix, $C_K: [S_{Pr}, S_{Po}, S_K, S_H]$, elaborates the sets for Presence, Power, Knowledge, and Harmony, each containing multiple elements. For example, as explored shortly in the section on the s-Group, various elements derived from the four sets define the behavior of s-Group atoms and could be functions such as 'power', 'energy', 'adventure', and 'courage', amongst others, hence collectively describing s-Groups atoms' way of being. Specifically, Line 5, $C_N: f(S_{Pr} \times S_{Po} \times S_K \times S_H)$, suggests that unique seeds are created from a combination of such elements from all four sets, with a particular element leading, that in effect creates the distinctness possible at the level of atoms.

Line 6, $\left(\downarrow R_{C_U} = f(R_{C_N})\right)$, specifies quantization between the layer where the seeds are formed, and the physical layer we are familiar with, and as explored in Chapter 3.5 and 3.6, will result in Line 7, $C_U: [P, V, M, C]$, hence changing the material-fabric of existence. The possibilities represented by Lines 1 through 5 hence concretize through the quantization represented by Line 6 to become the atoms with its physical (related to Presence), vital (related to Power), mental (related to Knowledge), and connection (related to Harmony) aspects now existing in material reality typified by Light moving at c. Note that just as Line 6 represents a process of quantization relating the layer of reality created by Light traveling at c with the antecedent layers, so too Lines 2 and 4 as previously discussed, also represent quantization of a more subtle kind that ultimately plays a critical part in allowing the material-fabric to express infinite diversity.

Typically it is the process as captured by the Space-Matrix that will determine if Line 6 is activated. Specifically patterns at the untransformed layer, U, will need to be overcome, as specified by the second-line from the bottom of the Space-Matrix: $(\uparrow > P_x)$. But as specified by the bottom-line of the Time-Matrix, reproduced below, it is only with the advent of the human-system that the automaticity of the action of meta-levels is reversed:

$$U \to \begin{array}{l} t \leq E_{Cell}; TC: M_3 \to U \\ t \sim E_{Human}; TC: U \to M_3 \end{array}$$

Hence in the case of atoms, which in this emergence is a pre-human system, the fact that patterns do not need to be overcome means that quantization happens automatically.

Nonetheless, and given this context it is useful to review
Equation 3.6.5, Quantization Effect of Organization on Material-Fabric:

$$Impact \ on \ Materal \ Fabric = \begin{bmatrix} |[L][S][T]TC \to x_T|_{(x_U | x_T)} \\ \times \\ (|mod(Z_Q)|_{Y > U}) \\ \ni \\ Z_Q, Z \in U \ (Space, Time, Energy, Gravity) \end{bmatrix}$$

Line 1 in the matrix is simply (3.6.1) the Simplified Light-Space-Time Emergence equation. To understand the quantization that may occur, the organization resulting from the application of Line 1 is multiplied $(\times)$ by a modulation (mod) of a fourfold space-time-energy-gravity quantization (Z_Q), so long as Y>U, that is, the operative bases are relatively transformed and hence have an active influence greater than U. The fourfold quantization is specified by $(\ni Z_Q, Z \in U \ (Space, Time, Energy, Gravity))$ where $(\ni)$ the quantization

144

being applied (Z) is each (U) of the elements of the set (Space, Time, Energy, Gravity).

But as just summarized in the Time-Matrix in (3.1.3) Y is by definition greater than U and hence quantization is automatic. In terms of atoms such quantization implies that wholeness becomes fully active through specific space, time, energy, and gravity quantization to create an holistic "ecosystem" with its own "atom logic" as it were. The wholeness has now precipitated into the material-fabric and is available to be consciously and unconsciously tapped into.

Precipitation of s-Group Logic into Material-Fabric

The s-Group consists primarily of what are known as alkali metals and alkali earth metals. These alkali metals are known to easily lose electrons and form what is known as positive ions. When they lose electrons energy is gained, but when the electrons are taken up by other atoms in proximity there is a lot of energy released. Some have referred to these groups as "violent worlds" (Tweed, 2003), and it has been pointed out that stars shine because they are transmuting vast amount of hydrogen into helium, both of which are s-Group elements. So one gets the sense that the s-Group may be an emergence of Light's property of Power.

Stars and suns are also known to be the crucibles where all the different kinds of atoms are created. So these furnaces of power by virtue of their heat and high pressure are able to force electrons and protons and neutrons to come together to create all the different types of atoms known in the universe.

But philosophically what are the s-Group atoms or elements? These are atoms where there is an equal likelihood of an electron being anywhere in a symmetrical sphere around the nucleus. The other groups are all similarly defined by likelihoods of electrons being within a possible pattern around the nucleus. The patterns that distinguishes s-Group atoms is a sphere, and since all other patterns can be thought of as occurring within some sphere, in some sense this is like an imprint or precipitation or emergence that allows other kinds of emergences to surface within it. So the elements that are part of the s-Group may be thought of as the adventurers with courage who venture into a brave new world to create some foundation by which all other element-creations can follow.

The fact that hydrogen and helium are known to constitute 98% of the Universe (Heiserman, 1991) relative to other elements therefore makes sense in this view, especially since hydrogen and helium provide the fuel with which the star-furnaces manufacture all other elements.

So s-Group elements seem to embody functions such as power, energy, adventure, courage, and can be thought of as an emergence of the property of Light to do with Power.

Hence, a series of equations linked to S_{System_P} as the prime set can be suggested, starting with the s-Group mapping, as in Equation 6.3.1:

$$Element_{s-Group} = Xa + Y\bar{b}_{0-n} \ where \left[\begin{array}{c} X \in [S_{System_P}] \\ Y \in [S_{System_{Pr}}, S_{System_P}, S_{System_K}, S_{System_N}] \\ a, b \ are \ integers; a > b \end{array} \right]$$

Eq 6.3.1: s-Group Element

Note that (6.3.1) already implies that Lines 1 – 5 in the Light Matrix (3.1.3) have been activated, and that the logic of the s-Group-ecosystem will automatically precipitate into the material-fabric through the action of Line 6-7 of (3.1.3).

Further, the equivalent mapping between traditional element groupings and S_{System_P} as in Equations 6.3.2 and 6.3.3 can be specified:

$$Element_{Alkali \ metal} =$$

$$Xa + Y\bar{b}_{0-n} \ where \left[\begin{array}{c} X \in [S_{System_P}] \\ Y \in [S_{System_{Pr}}, S_{System_P}, S_{System_K}, S_{System_N}] \\ a, b \ are \ integers; a > b \end{array} \right]$$

Eq 6.3.2: Alkali Metal Element

$$Element_{Alkali \ earth \ metal} =$$

$$Xa + Yb^-_{0-n} \quad where \left[\begin{array}{c} X \in [S_{System_P}] \\ Y \in [S_{System_{P'}}, S_{System_P}, S_{System_K}, S_{System_N}] \\ a, b \ are \ integers; a > b \end{array} \right]$$

Eq 6.3.3: Alkali Earth Metal Element

Further, as a representative element belonging to the Alkali Metal group the equation for Lithium (Li), as in Equation 6.3.4, would be:

$$Element_{Lithium} = Xa + Yb^-_{0-n} \quad where \left[\begin{array}{c} X \in [S_{System_P}] \\ Y \in [S_{System_{P'}}, S_{System_P}, S_{System_K}, S_{System_N}] \\ a, b \ are \ integers; a > b \end{array} \right]$$

Eq 6.3.4: Lithium

Note that the primary element X in all these cases may be a function or attribute along the lines of 'expresses power'. The secondary elements Y in all these would have some elements being the same, and then would have specific differences as one gets into sub-classes and the individual elements themselves.

Precipitation of p-Group Logic into Material-Fabric

Atoms or elements belonging to the p-Group are those with the likelihood of electrons occurring equally on either side of the nucleus, like a dumbbell of sorts.

There are some very significant elements in this group that are part of the metal, metalloid, non-metal, halogen, and noble gas sub-groupings. Carbon, Nitrogen, Oxygen, and Silicon are some of the sample elements. Looking at the types of elements present in this group it is as though all the element possibilities have been represented within it. It is perhaps that the possibility of ideas behind all elements has emerged in this group and one can hypothesize that this group may be a reflection of the property of Knowledge, forming archetypes from which all other elements are created.

Philosophically, the one spherical shell or probability cloud (s) becoming two shells like a dumbbell (p) signifies the creation of an essential polarity within a unit space. So in a sense the one becoming two is the first instance of variability in space. The two, spaced in 3-dimensions around the nucleus in a sense create six switches becoming an attractor or allowing a vaster number of different kinds of elements to surface. So there is a sense of the 'idea' of the element that comes into focus.

But further, the essential elements that allow both thinking and virtual thinking machines to come into being, are also contained within this group. Carbon is the basis of DNA and of all life. The fact the Silicon, directly below it in the periodic table and therefore sharing essential qualities, is considered the basis of all virtual thinking machines is therefore perhaps significant and may reinforce the notion that the p-Group is a precipitation of Light's property of Knowledge.

Hence, a series of equations linked to S_{System_K} as the prime set can be suggested, starting with the p-Group mapping, as in Equation 6.3.5:

$$Element_{\,p-Group} = Xa + Yb_{0-n}^{-} \;\; where \left[\begin{array}{c} X \in [S_{System_K}] \\ Y \in [S_{System_{Pr}}, S_{System_p}, S_{System_{K'}}, S_{System_N}] \\ a, b \;are\; integers; a > b \end{array} \right]$$

Eq 6.3.5: p-Group Element

Note that (6.3.5) already implies that Lines 1 – 5 in the Light Matrix (3.1.3) have been activated, and that the logic of the p-Group-ecosystem will automatically precipitate into the material-fabric through the action of Line 6-7 of (3.1.3).

Further, the equivalent mapping between traditional element groupings and S_{System_K} as in Equations 6.3.6 through 6.3.11 can also be specified:

$$Element_{\,Metal} = Xa + Yb_{0-n}^{-} \;\; where \left[\begin{array}{c} X \in [S_{System_K}] \\ Y \in [S_{System_{Pr}}, S_{System_p}, S_{System_{K'}}, S_{System_N}] \\ a, b \;are\; integers; a > b \end{array} \right]$$

Eq 6.3.6: Metal Element

$$Element_{\,Metalloid} = Xa + Yb_{0-n}^{-} \;\; where \left[\begin{array}{c} X \in [S_{System_K}] \\ Y \in [S_{System_{Pr}}, S_{System_p}, S_{System_{K'}}, S_{System_N}] \\ a, b \;are\; integers; a > b \end{array} \right]$$

Eq 6.3.7: Metalloid Element

$$Element_{Non-Metal} = Xa + Y\bar{b}_{0-n} \quad where \begin{bmatrix} X \in [S_{System_K}] \\ Y \in [S_{System_{Pr}}, S_{System_p}, S_{System_K}, S_{System_N}] \\ a, b \ are \ integers; a > b \end{bmatrix}$$

Eq 6.3.8: Non-Metal Element

$$Element_{Halogen} = Xa + Y\bar{b}_{0-n} \quad where \begin{bmatrix} X \in [S_{System_K}] \\ Y \in [S_{System_{Pr}}, S_{System_p}, S_{System_K}, S_{System_N}] \\ a, b \ are \ integers; a > b \end{bmatrix}$$

Eq 6.3.9: Halogen Element

$$Element_{Noble\ Gas} = Xa + Y\bar{b}_{0-n} \quad where \begin{bmatrix} X \in [S_{System_K}] \\ Y \in [S_{System_{Pr}}, S_{System_p}, S_{System_K}, S_{System_N}] \\ a, b \ are \ integers; a > b \end{bmatrix}$$

Eq 6.3.10: Noble Gas Element

Further, as a representative element belonging to the Non-Metal group the equation for Carbon (C), as in Equation 6.3.11, would be:

$$Element_{Carbon} = Xa + Y\bar{b}_{0-n} \quad where \begin{bmatrix} X \in [S_{System_K}] \\ Y \in [S_{System_{Pr}}, S_{System_p}, S_{System_K}, S_{System_N}] \\ a, b \ are \ integers; a > b \end{bmatrix}$$

Eq 6.3.11: Carbon

Note that the primary element X in all these cases may be a function or attribute along the lines of 'knowledge'. The secondary elements Y in all these would have some elements being the same, and then would have specific differences as one gets into sub-classes and the individual elements themselves.

Precipitation of d-Group Logic into Material-Fabric

The d-Group comprises the Transition Metals. These metals are generally hard and strong, exhibit corrosive resistance, and can be thought of as workhorse elements. Many industrial and well-known elements sit in this group: Titanium,

149

Chromium, Manganese, Iron, Cobalt, Nickel, Copper, Zinc, Silver, Platinum, and Gold, amongst others.

The elements and atoms in the d-Group are equally likely to show up in four possible lobes or probability-spaces around the nucleus. Four lobes occurring in multiple possible planes around the nucleus will likely create a space of stability, since there is a possibility of four lobes creating the four vertices of a tetrahedron (Fuller, 1982) that has been positioned as one of the most stable shapes in the universe.

Much of the constructed world around us is created from these elements. Further, most of the series in the group easily lose one or more electrons thereby easily combining with other atoms to form a vast array of compounds. Also, looking more broadly at the function of these elements, it can be seen that these metals exist for service, to help bring about perfection in the constructed world, to help much of the machinery in which they are used, and to assist the processes dependent on them to be completed with diligence. Hence, these transition metals appear to be an emergence of Light's property of Presence.

Therefore, a series of equations linked to $S_{System_{Pr}}$ as the prime set can be suggested, starting with the d-Group mapping, as in Equation 6.3.12:

$$Element_{d-Group} = Xa + Y\bar{b}_{0-n} \quad where \begin{bmatrix} X \in [S_{System_{Pr}}] \\ Y \in [S_{System_{Pr}}, S_{System_{p}}, S_{System_{K}}, S_{System_{N}}] \\ a, b \ are \ integers; a > b \end{bmatrix}$$

Eq 6.3.12: d-Group Element

Note that (6.3.12) already implies that Lines 1 – 5 in the Light Matrix (3.1.3) have been activated, and that the logic of the d-Group-ecosystem will automatically precipitate into the material-fabric through the action of Line 6-7 of (3.1.3).

Further, the equivalent mapping between traditional element groupings and $S_{System_{Pr}}$, as in Equation 6.3.13, can also be specified:

$$Element_{Transition \ Metal} =$$

$$Xa + Yb^-_{0-n} \quad where \begin{bmatrix} X \in [S_{System_{Pr}}] \\ Y \in [S_{System_{Pr}}, S_{System_{P}}, S_{System_{K}}, S_{System_{N}}] \\ a, b \ are \ integers; a > b \end{bmatrix}$$

Eq 6.3.13: Transition Metal Element

Further, as a representative element belonging to the Transition-Metal group the equation for Gold (Au), as in Equation 6.3.14, would be:

$$Element_{\ Gold} = Xa + Yb^-_{0-n} \quad where \begin{bmatrix} X \in [S_{System_{Pr}}] \\ Y \in [S_{System_{Pr}}, S_{System_{P}}, S_{System_{K}}, S_{System_{N}}] \\ a, b \ are \ integers; a > b \end{bmatrix}$$

Eq 6.3.14: Gold

Note that the primary element X in all these cases may be a function or attribute along the lines of 'work horse'. The secondary elements Y in all these would have some elements being the same, and then would have specific differences as one gets into sub-classes and the individual elements themselves.

Precipitation of f-Group Logic into Material-Fabric

The f-Group comprises of the Lanthanides and Actinides. Philosophically, elements in the f-Group consist of 6 lobes around the nucleus within which an electron may be found. 6 lobes will exist in multiple planes around the nucleus and suggests the notion of extended relationship and collectivity: the attempt to build larger and larger bonds within a small space. Considering this it is likely that the f-Group is an emergence of Light's property of Harmony.

Thinking about Lanthanides, some interesting facts may reinforce this notion. First, the spin of electrons in a lanthanides' outer shell is aligned, creating a strong magnetic field. The notion of creating a strong magnetic field seems to be consistent with the notion of engendering a collectivity through the ordered attraction and repulsion of elements. Second, these elements curiously occur together in nature often in the same ores (Gray, 2009) and are chemically interchangeable also suggesting the notion of forming a tight intra-group collectivity.

Actinides on the other hand are inherently radioactive. This implies that these elements have inherently crossed a threshold of stability and have the urge, over their own half-lives, to decompose into other elements. This natural urge may suggest some boundary conditions on the notion of collectivity and nurturing, giving additional insight into the nature of collectivity and nurturing. Further, the entire actinide group, as opposed to the lanthanide group that is inherently stable, is unstable. It is curious that both these should be part of the f-Group, and they must provide insight into boundary conditions into the notion of collectivity in elements.

Hence, a series of equations linked to S_{System_N} as the prime set can be suggested, starting with the f-Group mapping, as in Equation 6.3.15:

$$Element_{f-Group} = Xa + Yb_{0-n}^{-} \quad where \left[\begin{array}{c} X \in [S_{System_N}] \\ Y \in [S_{System_{Pr}}, S_{System_P}, S_{System_K}, S_{System_N}] \\ a, b \; are \; integers; a > b \end{array} \right]$$

Eq 6.3.15: f-Group Element

Note that (6.3.15) already implies that Lines 1 – 5 in the Light Matrix (3.1.3) have been activated, and that the logic of the f-Group-ecosystem will automatically precipitate into the material-fabric through the action of Line 6-7 of (3.1.3).

Further, the equivalent mapping between traditional element groupings and S_{System_N}, as in Equations 6.3.16 and 6.3.17, can also be specified:

$$Element_{Lanthanide} = Xa + Yb_{0-n}^{-} \quad where \left[\begin{array}{c} X \in [S_{System_N}] \\ Y \in [S_{System_{Pr}}, S_{System_P}, S_{System_K}, S_{system_N}] \\ a, b \; are \; integers; a > b \end{array} \right]$$

Eq 6.3.16: Lanthanide Element

$$Element_{Actinide} = Xa + Yb_{0-n}^{-} \quad where \left[\begin{array}{c} X \in [S_{System_N}] \\ Y \in [S_{System_{Pr}}, S_{System_P}, S_{System_K}, S_{System_N}] \\ a, b \; are \; integers; a > b \end{array} \right]$$

Eq 6.3.17: Actinide Element

Further, as a representative element belonging to the Lanthanide group the equation for Lanthanum (La), as in Equation 6.3.18, would be:

$$Element_{Lanthanum} = Xa + Yb_{0-n}^{-} \quad where \begin{bmatrix} X \in [S_{System_N}] \\ Y \in [S_{System_{Pr}}, S_{System_P}, S_{System_K}, S_{System_N}] \\ a, b \ are \ integers; a > b \end{bmatrix}$$

Eq 6.3.18: Lanthanum

Note that the primary element X in all these cases may be a function or attribute along the lines of 'experiments in collectivity'. The secondary elements Y in all these would have some elements being the same, and then would have specific differences as one gets into sub-classes and the individual elements themselves.

SECTION 7: COMPUTING LIFE

This section explores the quantum-level computation that causes the emergence of life through the cell, complex human attributes such as thoughts and feelings, and uniqueness of individuality.

Hence Chapter 7.1 will explore the quantum-level computation in the emergence and complexification of four-foldness through the primary molecular plans of nucleic acids, proteins, lipids, and polysaccharides.

Chapter 7.2 will relate key human attributes of sensations, urges, desires, wills, feelings, emotions, and thought to the continuing journey of fourfold complexification.

Chapter 7.3 will relate truer individuality to the fourfold properties implicit in Light.

Chapter 7.1: Quantum-Level Computation in Creation of a Living Cells Ecosystem

The living cell has a universe of adaptability embedded in it. In 'The Machinery of Life', Goodsell, an Associate Professor of Molecular Biology at the Scripps Research Institute (Goodsell, 2010) suggests that every living thing on Earth uses a similar set of molecules to eat, to breathe, to move, and to reproduce. There are molecular machines that do the myriad things that distinguish living organisms that are identical in all living cells. This nanoscale machinery of cells uses four basic molecular plans with unique chemical personalities: nucleic acids, proteins, lipids, and polysaccharides.

In this chapter we will review the quantum-level computation that facilitates Light's emergence as Living Cells,

Light's Emergence as Living Cells

As discussed previously the Light-Space-Time Emergence equation (3.1.3) being iterative, can be used to model emergence as it proceeds from simpler four-fold to more complex four-fold manifestations. Hence (3.1.3) has already been applied to suggest the emergence of the electromagnetic spectrum, quantum particles in general, bosons as a further instance of a particular kind of quantum particle, and atoms. Here it will be applied to suggest the emergence of Living Cells. But further, as implied by (3.6.5), the Quantization Effect of Organization on Material-Fabric equation, any process of deeper organization, such as is responsible for the architecture and cohesiveness of living cells, has the possibility of altering the material-fabric or fabric of existence so long as the bases involved are driven primarily by a meta-level.

As suggested by (3.1.3), reproduced below for convenience, the architecture and details of living cells can be seen to be the result of the application of the Light, Space, and Time matrices as will be elaborated:

$Emergence_{light-space-time} =$

$$\left| \begin{array}{c} \begin{vmatrix} C_\infty : [Pr, Po, K, H] \\ \left(\downarrow R_{C_K} = f(R_{C_\infty}) \right) \\ C_K : [S_{Pr}, S_{Po}, S_K, S_H] \\ \left(\downarrow R_{C_N} = f(R_{C_K}) \right) \\ C_N : f(S_{Pr} \times S_{Po} \times S_K \times S_H) \\ \left(\downarrow R_{C_U} = f(R_{C_N}) \right) \\ C_U : [P,V,M,C] \end{vmatrix} Light \quad \begin{vmatrix} M_3 \to System_X \\ (\uparrow F \to I) \\ M_2 \to S_{System_X} \\ (\uparrow Sig \to F) \\ M_1 \to Sig_x \\ (\uparrow > P_x) \\ U \to x_U \end{vmatrix} Space \\[2mm] \begin{vmatrix} M_3 : -\infty \le t \le \infty \\ \downarrow \\ M_2 : 0 \ge t > \infty \\ \downarrow \\ M_1 : 0 > t > \infty \quad Time \\ \downarrow \\ U \to \begin{array}{l} t \le E_{Cell}; TC: M_3 \to U \\ t \sim E_{Human}; TC: U \to M_3 \end{array} \end{vmatrix} \begin{array}{c} \langle x_U | x_T \rangle \\[30mm] TC \to x_T \end{array} \end{array} \right|$$

Starting with the Light-Matrix, the top left-hand matrix in (3.1.3), the first line from the top, $C_\infty : [Pr, Po, K, H]$, specifies the fundamental architecture of living cells. As will be explored shortly, Nucleic Acids are an emergence of Light's property of Knowledge, Polysaccharides are an emergence of Light's property of Power, Proteins are an emergence of Light's property of Presence, and Lipids are an emergence of Light's property of Harmony. The fundamental architecture of these aspects hence, is an emergence of the properties of Light at ∞.

Line 3 in the Light-Matrix, $C_K : [S_{Pr}, S_{Po}, S_K, S_H]$, elaborates the sets for Presence, Power, Knowledge, and Harmony, each containing multiple elements. For example, as will be explored in the section on Proteins, various elements derived from the four sets define the behavior of Proteins and could be functions such as 'exist for service', 'to bring about perfection at the level of the cell', 'extreme diligence and perseverance', amongst others, hence collectively describing Proteins' way of being. Specifically, Line 5, $C_N : f(S_{Pr} \times S_{Po} \times S_K \times S_H)$, suggests that unique seeds are created from a combination of such elements from all four sets, with a particular element leading, that in effect creates the distinctness possible at the level of cells.

Line 6, $\left(\downarrow R_{C_U} = f(R_{C_N}) \right)$, specifies quantization between the layer where the seeds are formed, and the physical layer we are familiar with, and as explored in Chapter 3.5 and 3.6, will result in Line 7, $C_U : [P,V,M,C]$, hence changing the material-fabric of existence. The possibilities represented by Lines 1 through 5 hence concretize through the quantization represented by Line 6 to become or enhance living cells with its subtle physical (related to Presence), vital (related to Power), mental (related to Knowledge), and connection (related to Harmony) aspects now existing in material reality typified by Light moving at c. Note that just as Line 6 represents a process of quantization relating the layer of reality created by Light traveling at c with the antecedent layers, so too Lines 2 and 4 as previously discussed, also represent quantization of a more subtle kind that ultimately plays a critical part in allowing the material-fabric to express infinite diversity.

Typically it is the process as captured by the Space-Matrix that will determine if Line 6 is activated. Specifically patterns at the untransformed layer, U, will need to be overcome, as specified by the second-line from the bottom of the Space-Matrix: $(\uparrow > P_x)$. But as specified by the bottom-line of the Time-Matrix, reproduced below, it is only with the advent of the human-system that the automaticity of the action of meta-levels is reversed:

$$U \rightarrow \begin{array}{l} t \le E_{Cell}; TC: M_3 \rightarrow U \\ t \sim E_{Human}; TC: U \rightarrow M_3 \end{array}$$

Hence in the case of living cells, which in this emergence is a pre-human system, the fact that patterns do not need to be overcome means that quantization more or less happens automatically.

Nonetheless, and given this context it is useful to review
Equation 3.6.5, Quantization Effect of Organization on Material-Fabric:

$$Impact\ on\ Materal\ Fabric = \begin{bmatrix} |[L][S][T]TC \rightarrow x_T|_{\langle x_U | x_T \rangle} \\ \times \\ (|mod(Z_Q)|_{Y > U}) \\ \ni \\ Z_Q, Z \in U\ (Space, Time, Energy, Gravity) \end{bmatrix}$$

Line 1 in the matrix is simply (3.6.1) the Simplified Light-Space-Time Emergence equation. To understand the quantization that may occur, the organization resulting from the application of Line 1 is multiplied $(\times)$ by a modulation (mod) of a fourfold space-time-energy-gravity quantization (Z_Q), so long as Y>U, that is, the operative bases are relatively transformed and hence have an active influence greater than U. The fourfold quantization is specified by $(\ni Z_Q, Z \in U \ (Space, Time, Energy, Gravity))$ where ($\ni$) the quantization being applied (Z) is each (U) of the elements of the set (Space, Time, Energy, Gravity).

But as just summarized in the Time-Matrix in (3.1.3) Y is by definition greater than U and hence quantization is automatic. In terms of living cells such quantization implies that wholeness becomes fully active through specific space, time, energy, and gravity quantization to create an holistic "ecosystem" with its own "living cell logic" as it were. The wholeness has now precipitated into the material-fabric and is available to be consciously and unconsciously tapped into.

Precipitation of Nucleic Acids Logic into Material-Fabric

Nucleic acids basically encode information. They store and transmit the genome, the hereditary information needed to keep the cell alive. They function as the cell's librarians and contain information on how to make proteins and when to make them.

They are hence, the keepers of a cell's knowledge, its wisdom, its ability to make laws, the vehicle to spread knowledge within cells and to the next generation of cells. Being so, one can see that there is similarity with the set for system-knowledge highlighted earlier in Chapter 2.3. Reproducing Equation 2.3.3:

$$S_{System_K} \ \ni \ [Wisdom, Law \ Making, Spread \ of \ Knowledge...]$$

Nucleic acids can therefore be thought of as a precipitation of system-knowledge at the cellular level.

Hence, a nucleic acid will have a generalized signature, as in Equation 7.1.1, derived from the system-knowledge family:

$$Sig_{nucleic \ acid} = Xa + Yb^-_{0-n} \ \ where \ \begin{bmatrix} X \in [S_{System_K}] \\ Y \in [S_{System_{Pr}}, S_{System_P}, S_{System_K}, S_{System_N}] \\ a, b \ are \ integers; a > b \end{bmatrix}$$

Eq 7.1.1: Nucleic Acid

The primary element X could be an attribute or function such as 'keeper of knowledge'. Secondary elements Y could be 'protein laws', 'generational knowledge', amongst others. The collectivity of elements as per the equation would specify the character of a nucleic acid.

Note that (7.1.1) already implies that Lines 1 – 5 in the Light Matrix (3.1.3) have been activated, and that the logic of the nucleic-acid-ecosystem will automatically precipitate into the material-fabric through the action of Line 6-7 of (3.1.3).

DNA and RNA, two types of nucleic acids, would hence have the equations as specified by Equation 7.1.2 and 7.1.3 respectively.

$$Sig_{DNA} = Xa + Yb_{0-n}^{-} \ where \left[\begin{array}{c} X \in [S_{System_K}] \\ Y \in [S_{System_{Pr}}, S_{System_P}, S_{System_K}, S_{System_N}] \\ a, b \ are \ integers; a > b \end{array} \right]$$

Eq 7.1.2: DNA

$$Sig_{RNA} = Xa + Yb_{0-n}^{-} \ where \left[\begin{array}{c} X \in [S_{System_K}] \\ Y \in [S_{System_{Pr}}, S_{System_P}, S_{System_K}, S_{System_N}] \\ a, b \ are \ integers; a > b \end{array} \right]$$

Eq 7.1.3: RNA

The primary element X in (7.1.2) and (7.1.3) would be the same as that for nucleic acids (7.1.1). The secondary elements Y however will be a larger and more specific set with many elements in common with (7.1.1).

Precipitation of Proteins Logic into Material-Fabric

Proteins are the cells work-horses. Look anywhere in a cell and one will see proteins at work. Proteins are built in thousands of shapes and sizes, each performing a different function. As Goodsell describes, "some are built simply to adopt a defined shape, assembling into rods, nets, hollow spheres, and tubes. Some are molecular motors, using energy to rotate, or flex, or crawl. Many are chemical catalysts that perform chemical reactions atom-by-atom, transferring and transforming chemical groups exactly as needed." With their wide potential for diversity, proteins are constructed to perform most of the everyday tasks of the cells. In fact human cells build around 30,000 different kinds of proteins to execute on the diverse array of cellular level tasks.

Proteins hence, exist for service, to bring about perfection at the level of the cell, are characterized by extreme diligence and perseverance, and so on. Being so, one can see that there is similarity with the set for system-presence highlighted earlier in Chapter 2.3. Reproducing Equation 2.3.1:

$$S_{System_{Pr}} \ni [Service, Perfection, Diligence, Perseverance, ...]$$

Proteins can therefore be thought of as a precipitation of system-presence at the cellular level.

Hence, a protein could have a generalized signature, as in Equation 7.1.4, derived from the system-presence family:

$$Sig_{protein} = Xa + Y\bar{b}_{0-n} \;\; where \begin{bmatrix} X \in [S_{System_{Pr}}] \\ Y \in [S_{System_{Pr}}, S_{System_p}, S_{System_K}, S_{System_N}] \\ a, b \; are \; integers; a > b \end{bmatrix}$$

Eq 7.1.4: Protein

This could yield a vast number of functional proteins. In fact it may be possible that the 30,000 or so known proteins created by the human cell could each be specified by a signature equation of this nature. It may be possible to map existing proteins to functionality as suggested by the four sets of molecular plans.

Note that (7.1.4) already implies that Lines 1 – 5 in the Light Matrix (3.1.3) have been activated, and that the logic of the protein-ecosystem will automatically precipitate into the material-fabric through the action of Line 6-7 of (3.1.3).

Consider Insulin, for example. Insulin regulates the metabolism of carbohydrates, fats and protein by promoting the absorption of, especially, glucose from the blood into fat, liver and skeletal muscle cells. Equation 7.1.5 for Insulin would hence be:

$$Sig_{insulin} = Xa + Y\bar{b}_{0-n} \;\; where \begin{bmatrix} X \in [S_{System_{Pr}}] \\ Y \in [S_{System_{Pr}}, S_{System_p}, S_{System_K}, S_{System_N}] \\ a, b \; are \; integers; a > b \end{bmatrix}$$

Eq 7.1.5: Insulin

The primary element X could be an attribute or function such as 'workhorse. Secondary elements Y could be 'metabolic regulation, 'glucose absorption', 'blood to fat channel', amongst others. The collectivity of elements as per the equation would specify the character of insulin.

Consider Histones as another example. They are the chief protein components of chromatin, acting as spools around which DNA winds, and playing a role in gene regulation. Without histones, the unwound DNA in chromosomes would be very long (a length to width ratio of more than 10 million to 1 in human DNA). Equation 7.1.6 for Histones would hence be:

$$Sig_{histones} = Xa + Yb_{0-n}^{-} \quad where \begin{bmatrix} X \in [S_{System_{Pr}}] \\ Y \in [S_{System_{Pr}}, S_{System_{P}}, S_{System_{K}}, S_{System_{N}}] \\ a, b \ are \ integers; a > b \end{bmatrix}$$

Eq 7.1.6: Histones

The secondary element Y would have elements such as 'gene regulation', 'unwound DNA management', amongst others.

Precipitation of Lipids Logic into Material-Fabric

Lipids by themselves are tiny molecules, but when grouped together form the largest structures of the cell. When placed in water, lipid molecules aggregate to form huge waterproof sheets. These sheets easily form boundaries at multiple levels and allow concentrated interactions and work to be performed within a cell. Hence, the nucleus and the mitochondria are contained within lipid-defined compartments. Similarly, each cell itself is contained within a lipid-defined boundary.

Lipids are therefore promoters of relationship, of harmony in the cell, of nurturing the cell-level division of labor, of allowing specialization and uniqueness to emerge, hence perhaps of earlier forms of compassion and love, and so on. The notion of such early forms of compassion is consistent with the biologist's perspective that at some point a gene for compassion was developed in pre-human species (Wright, 2009). Being so, one can see that there is similarity with the set for system-nurturing highlighted earlier in Chapter 2.3. Reproducing Equation 2.3.4:

$$S_{System_{N}} \ni [Love, Compassion, Harmony, Relationship ...]$$

This function of harmonization suggests that lipids can therefore be thought of as a precipitation of system-nurturing at the cellular level.

Lipids could have a generalized signature, as in Equation 7.1.7, derived from the system-nurturing family:

$$Sig_{lipid} = Xa + Yb_{0-n}^{-} \quad where \begin{bmatrix} X \in [S_{System_N}] \\ Y \in [S_{System_{Pr}}, S_{System_P}, S_{System_K}, S_{System_N}] \\ a, b \ are \ integers; a > b \end{bmatrix}$$

Eq 7.1.7: Lipid

Note that (7.1.7) already implies that Lines 1 – 5 in the Light Matrix (3.1.3) have been activated, and that the logic of the lipid-ecosystem will automatically precipitate into the material-fabric through the action of Line 6-7 of (3.1.3).

Specific lipids such as monoglycerides and phospholipids cold have the following equations:

$$Sig_{monoglyceride} = Xa + Yb_{0-n}^{-} \quad where \begin{bmatrix} X \in [S_{System_N}] \\ Y \in [S_{System_{Pr}}, S_{System_P}, S_{System_K}, S_{System_N}] \\ a, b \ are \ integers; a > b \end{bmatrix}$$

Eq 7.1.8: Monoglyceride

$$Sig_{phospholipids} = Xa + Yb_{0-n}^{-} \quad where \begin{bmatrix} X \in [S_{System_N}] \\ Y \in [S_{System_{Pr}}, S_{System_P}, S_{System_K}, S_{System_N}] \\ a, b \ are \ integers; a > b \end{bmatrix}$$

Eq 7.1.9: Phospholipids

The primary element X shared by each of the lipids could be an attribute or function such as 'compartmentalization'. Secondary elements Y could be of the nature of 'work breakdown', 'intra-cell love', amongst others, and would vary with each different kind of lipid.

Precipitation of Polysaccharides Logic into Material-Fabric

Polysaccharides are long, often branched chains of sugar molecules. Sugars are covered with hydroxyl groups, which associate to form storage containers. As a

162

result polysaccharides function as the storehouse of cell's energy. In addition polysaccharides are also used to build some of the most durable biological structures. The stiff shell of insects, for example are made of long polysaccharides.

Polysaccharides function to create energy, power, courage, strength thereby readying the cell for adventure, and so on. Being so, one can see that there is similarity with the set for system-power highlighted previously in Chapter 2.3. Reproducing Equation 2.3.2:

$$S_{System_p} \ni [Power, Courage, Adventure, Justice, ...]$$

Providing energy and strength, polysaccharides can be thought of as a precipitation of system-power at the cellular level.

Polysaccharides could have a generalized signature, as in Equation 7.1.10, derived from the system-power family:

$$Sig_{polysaccharide} = Xa + Yb_{0-n}^{-} \quad where \begin{bmatrix} X \in [S_{System_p}] \\ Y \in [S_{System_{Pr}}, S_{System_p}, S_{System_K}, S_{System_N}] \\ a, b \ are \ integers; a > b \end{bmatrix}$$

Eq 7.1.10: Polysaccharide

Note that (7.1.10) already implies that Lines 1 – 5 in the Light Matrix (3.1.3) have been activated, and that the logic of the polysaccharide-ecosystem will automatically precipitate into the material-fabric through the action of Line 6-7 of (3.1.3).

Glycogen is an example of a polysaccharide. Glycogen forms an energy reserve that can be quickly mobilized to meet a sudden need for glucose, but one that is less compact and more immediately available as an energy reserve than say triglycerides. Equation 7.1.11 for Glycogen follows:

$$Sig_{glycogen} = Xa + Yb_{0-n}^{-} \quad where \begin{bmatrix} X \in [S_{System_p}] \\ Y \in [S_{System_{Pr}}, S_{System_p}, S_{System_K}, S_{System_N}] \\ a, b \ are \ integers; a > b \end{bmatrix}$$

Eq 7.1.11: Glycogen

Cellulose is another example of a polysaccharide. Cellulose is a polymer made with repeated glucose units bonded together by beta-linkages. Humans and many animals lack an enzyme to break the beta-linkages, so they do not digest cellulose. Equation 7.1.12 for Cellulose follows:

$$Sig_{cellulose} = Xa + Y\bar{b}_{0-n} \quad where \begin{bmatrix} X \in [S_{System_P}] \\ Y \in [S_{System_{Pr}}, S_{System_P}, S_{System_K}, S_{System_N}] \\ a, b \ are \ integers; a > b \end{bmatrix}$$

Eq 7.1.12: Cellulose

The primary element for the preceding polysaccharides X would be along the lines of 'energy storage'. The secondary elements may vary, with a Y element for Glycogen being 'rapid energy deployment' for example, and a Y element for Cellulose being 'bonded energy', for example.

Chapter 7.2: Quantum-Level Computation in Creation of Sensations, Urges, Feelings, Thoughts Ecosystems

As human beings we experience sensations, urges and desires and wills, feelings and emotions, and thought. These are key aspects of our being and becoming and critical aspects of how choice at both the individual and collective levels may be determined.

This chapter will go over the quantum-level computation that suggests how light emerges as these capacities, and further, how these precipitate into the material-fabric.

Light's Emergence as Fundamental Capacities of Self

As discussed previously the Light-Space-Time Emergence equation (3.1.3) being iterative, can be used to model emergence as it proceeds from simpler four-fold to more complex four-fold manifestations. Hence (3.1.3) has already been applied to suggest the emergence of the electromagnetic spectrum, quantum particles in general, bosons as a further instance of a particular kind of quantum particle, atoms, and living cells. Here it will be applied to suggest the emergence of fundamental capacities of self. But further, as implied by (3.6.5), the Quantization Effect of Organization on Material-Fabric equation, any process of deeper organization, such as is responsible for the architecture and cohesiveness of capacities of self, has the possibility of altering the material-fabric or fabric of existence so long as the bases involved are driven primarily by a meta-level.

As suggested by (3.1.3), reproduced below for convenience, the architecture and details of capacities of self can be seen to be the result of the application of the Light, Space, and Time matrices as will be elaborated:

$$Emergence_{light-space-time} =$$

$$\begin{Vmatrix} \begin{bmatrix} C_\infty:[Pr, Po, K, H] \\ (\downarrow R_{C_K} = f(R_{C_\infty})) \\ C_K: [S_{Pr}, S_{Po}, S_K, S_H] \\ (\downarrow R_{C_N} = f(R_{C_K})) \\ C_N: f(S_{Pr} \times S_{Po} \times S_K \times S_H) \\ (\downarrow R_{C_U} = f(R_{C_N})) \\ C_U: [P,V,M,C] \end{bmatrix}_{Light} & \begin{bmatrix} M_3 \to System_X \\ (\uparrow F \to I) \\ M_2 \to S_{System_X} \\ (\uparrow Sig \to F) \\ M_1 \to Sig_X \\ (\uparrow > P_x) \\ U \to x_U \end{bmatrix}_{Space} \\ \begin{bmatrix} M_3: -\infty \leq t \leq \infty \\ \downarrow \\ M_2: 0 \geq t > \infty \\ \downarrow \\ M_1: 0 > t > \infty \\ \downarrow \\ U \to \begin{array}{l} t \leq E_{Cell}; TC: M_3 \to U \\ t \sim E_{Human}; TC: U \to M_3 \end{array} \end{bmatrix}_{Time} \; TC \to x_T & \langle x_U | x_T \rangle \end{Vmatrix}$$

Starting with the Light-Matrix, the top left-hand matrix in (3.1.3), the first line from the top, $C_\infty:[Pr, Po, K, H]$, specifies the architecture of the fundamental capacities of self to be introduced in this chapter. Hence, Thoughts are an emergence of Light's property of Knowledge, Urges, Desires, and Wills are an emergence of Light's property of Power, Sensations are an emergence of Light's property of Presence, and Feelings and Emotions are an emergence of Light's property of Harmony. The fundamental architecture of these aspects hence, is an emergence of the properties of Light at ∞.

Line 3 in the Light-Matrix, $C_K: [S_{Pr}, S_{Po}, S_K, S_H]$, elaborates the sets for Presence, Power, Knowledge, and Harmony, each containing multiple elements. For example, as will be explored in the section on Sensations, various elements derived from the four sets define the behavior of Sensations and could be functions such as 'tangible', 'take notice of', amongst others, hence collectively describing Sensations' way of being. Specifically, Line 5, $C_N: f(S_{Pr} \times S_{Po} \times S_K \times S_H)$, suggests that unique seeds are created from a combination of such elements from all four sets, with a particular element leading, that in effect creates the distinctness possible at the level of fundamental capacities of self.

Line 6, $(\downarrow R_{C_U} = f(R_{C_N}))$, specifies quantization between the layer where the seeds are formed, and the physical layer we are familiar with, and as explored in Chapter 3.5 and 3.6, will result in Line 7, $C_U: [P,V,M,C]$, hence changing the material-fabric of existence. The possibilities represented by Lines 1 through 5

166

hence concretize through the quantization represented by Line 6 to become or further enhance the capacities of self with its subtle physical (related to Presence), vital (related to Power), mental (related to Knowledge), and connection (related to Harmony) aspects now existing in material reality typified by Light moving at c. Note that just as Line 6 represents a process of quantization relating the layer of reality created by Light traveling at c with the antecedent layers, so too Lines 2 and 4 as previously discussed, also represent quantization of a more subtle kind that ultimately plays a critical part in allowing the material-fabric to express infinite diversity.

Typically it is the process as captured by the Space-Matrix that will determine if Line 6 is activated. Specifically patterns at the untransformed layer, U, will need to be overcome, as specified by the second-line from the bottom of the Space-Matrix: $(\uparrow > P_x)$. But as specified by the bottom-line of the Time-Matrix, reproduced below, it is only with the advent of the human-system that the automaticity of the action of meta-levels is reversed:

$$U \rightarrow \begin{matrix} t \leq E_{Cell}; TC: M_3 \rightarrow U \\ t \sim E_{Human}; TC: U \rightarrow M_3 \end{matrix}$$

Hence in the case of sensations, which in this emergence is largely a post-human system, the fact that patterns do need to be overcome means that quantization requires effort to happen. Given this, it is useful to review **Equation 3.6.5, Quantization Effect of Organization on Material-Fabric:**

$$Impact\ on\ Materal\ Fabric = \begin{bmatrix} |[L][S][T]TC \rightarrow x_T|_{\langle x_U | x_T \rangle} \\ \times \\ (|mod(Z_Q)|_{Y > U}) \\ \ni \\ Z_Q, Z \in U\ (Space, Time, Energy, Gravity) \end{bmatrix}$$

Line 1 in the matrix is simply (3.6.1) the Simplified Light-Space-Time Emergence equation. To understand the quantization that may occur, the organization resulting from the application of Line 1 is multiplied $(\times)$ by a modulation (mod) of a fourfold space-time-energy-gravity quantization (Z_Q), so long as Y>U, that is, the operative bases are relatively transformed and hence have an active influence greater than U. The fourfold quantization is specified by $(\ni Z_Q, Z \in U\ (Space, Time, Energy, Gravity))$ where $(\ni)$ the quantization being applied (Z) is each $(^U)$ of the elements of the set (Space, Time, Energy, Gravity).

But as just summarized in the Time-Matrix in (3.1.3) Y is by definition not greater than U and hence quantization is not automatic. In terms of Sensations such quantization implies that increasing wholeness can become fully active through specific space, time, energy, and gravity quantization to create an holistic "ecosystem" with its own "sensation logic" as it were. The wholeness only then precipitates into the material-fabric and is available to be consciously and unconsciously tapped into.

Precipitation of Sensations Logic into Material-Fabric

Sensations are those things we experience with our senses. We see things, hear things, and smell things, taste things, can touch things. This ability to enter into relationship with objects through sensation is nothing other than a result of the emergence of Light's property of Presence. We become present to Presence through the device of sensation. Sensation can be thought of as the means by which this property of Light – Presence - molds or ingrains itself in us as human beings. Its potentiality, all which is contained in this aspect of Light, becomes available to us through the power of sensation. Hence an equation, Equation 7.2.1, will generally represent the family of sensations. Some elements that it would comprise of may be 'tangible', 'take notice of', amongst others.

$$Sig_{sensation} = Xa + Yb^-_{0-n} \quad where \begin{bmatrix} X \in [S_{System_{Pr}}] \\ Y \in [S_{System_{Pr}}, S_{System_p}, S_{System_K}, S_{System_N}] \\ a, b \ are \ integers; a > b \end{bmatrix}$$

Eq 7.2.1: Sensation

There could also be equations for hearing, seeing, tasting, touching, and smelling.

But there is also a deeper experience of sensation that is possible. When we see things, for instance, what are we seeing? Is it just the surface rendering of the play of matter, or do we see that the fullness of Light is still there, with all its potentiality and possibility, in the smallest thing we look at? Do we see that the whole universe and more is present in all its fullness in the least thing that we easily ignore, or belittle, or loathe? When we touch things is it the seeming concreteness of the play of the particles or atoms or chains of molecules that we touch? Or is it the Love and Light and the vastness of all that IS that allows itself to be as a small corner that we touch so as to make infinity be felt by something so finite?

Such a deeper contact offered through sensation suggests a subset of (7.2.1) with secondary elements perhaps described as 'fullness of Light', 'contacting infinity', amongst others, thus also yielding an equation form (7.2.2):

$$Sig_{deeper-sensation} = Xa + Y\overline{b}_{0-n} \quad where \left[\begin{array}{c} X \in [S_{System_{Pr}}] \\ Y \in [S_{System_{Pr}}, S_{System_{P}}, S_{System_{K}}, S_{System_{N}}] \\ a, b \ are \ integers; a > b \end{array} \right]$$

Eq 7.2.2: Deeper Sensation

Note that (7.2.1 and 7.2.2) already implies that Lines 1 – 5 in the Light Matrix (3.1.3) have been activated, and that the logic of the sensations-ecosystem will automatically precipitate into the material-fabric through the action of Line 6-7 of (3.1.3).

Precipitation of Urges, Desires & Wills Logic into Material-Fabric

Urges and desires and wills are similarly a play of the emergence of Light's property of Power. In the mystery of focus, the vastness of Light has projected itself in us into an apparent smallness that is in reality everything that is. And this smallness is trying through urge and desire and will to connect viscerally or even intentionally to other smallnesses that similarly are nothing other than the fullness of Light projected into a small smorgasbord of selected function. So the urge or desire for food, or companionship, or of possession, or of climbing a peak, is nothing other than Light's compressed property of Power, trying to reach more of the fullness that it is through a fulfillment of the urge or desire or will that it masquerades as. Hence urges can be represented as Equation 7.2.3:

$$Sig_{urges} = Xa + Y\overline{b}_{0-n} \quad where \left[\begin{array}{c} X \in [S_{System_{P}}] \\ Y \in [S_{System_{Pr}}, S_{System_{P}}, S_{System_{K}}, S_{System_{N}}] \\ a, b \ are \ integers; a > b \end{array} \right]$$

Eq 7.2.3: Urges

Elements may be of the type of 'grasp', 'possess', 'deeply connect', amongst others.

169

Note that (7.2.3) already implies that Lines 1 – 5 in the Light Matrix (3.1.3) have been activated, and that the logic of the urges-ecosystem will automatically precipitate into the material-fabric through the action of Line 6-7 of (3.1.3).

Precipitation of Feelings & Emotions Logic into Material-Fabric

Feelings and emotions are a play of the emergence of Light's property of Harmony or Nurturing. Its instrument is the Heart and it generates an array of emotions that are an indication or active radar of whether we are moving toward or away from a reality of harmony, whether based on our small self or some larger Self of Light. Gradually, by navigating with these emotions and feelings we can get to a state where we always feel positive emotions which basically means we have more truly entered into relationship with some larger continent of Light. An equation for feelings is as represented by Equation 7.2.4:

$$Sig_{feelings} = Xa + Yb_{0-n}^{-} \quad where \begin{bmatrix} X \in [S_{System_N}] \\ Y \in [S_{System_{Pr}}, S_{System_P}, S_{System_K}, S_{System_N}] \\ a, b \text{ are integers}; a > b \end{bmatrix}$$

Eq 7.2.4: Feelings

Note that (7.2.4) already implies that Lines 1 – 5 in the Light Matrix (3.1.3) have been activated, and that the logic of the feelings-ecosystem will automatically precipitate into the material-fabric through the action of Line 6-7 of (3.1.3).

Precipitation of Thought Logic into Material-Fabric

Thoughts are a play of the emergence of Light's property of Knowledge. Through the thought we can become greater or conceptualize things greater or begun to enter into relationship with some things other than our small self. Thought allows us to connect to more "othernesses" or even the oneness of the reality of Light.

$$Sig_{thoughts} = Xa + Yb_{0-n}^{-} \quad where \begin{bmatrix} X \in [S_{System_K}] \\ Y \in [S_{System_{Pr}}, S_{System_P}, S_{System_K}, S_{System_N}] \\ a, b \text{ are integers}; a > b \end{bmatrix}$$

Eq 7.2.5: Thoughts

Note that (7.2.5) already implies that Lines 1 – 5 in the Light Matrix (3.1.3) have been activated, and that the logic of the thought-ecosystem will automatically precipitate into the material-fabric through the action of Line 6-7 of (3.1.3).

So we can see that the very substance of our becoming – sensations, urges, wills, desires, feelings, emotions, thoughts – are nothing other than the quantum-level computations of the four properties of light of presence, power, knowledge, and harmony.

Chapter 7.3: Quantum-Level Computation in Creation of Truer Individuality Ecosystem

At the core individuals can be thought of as projections of seeds formed in a vast continent of Light. So that core must always be there but it is often covered by surface dynamics of the physical, the vital, and the mental. These too are formed from properties of Light, but since the light has been separated from its source these movements and dynamics are incomplete in their nature.

It is easy for these dynamics and movements to completely occupy all individual processing power. This can happen to such an extent that the individual acting on the surface can become subject to these surface movements. But so long as there is the remembrance that even in separation and smallness - whether selfishness, myopia, fundamentalism, or any other ism - these movements are still light, then that can become the means by which the apparent chains around individuals can be loosened.

For in its essence these small movements are trying to extend into something other than what they are. By connecting the light in them with the larger Light this extension can yield something different than what is possible through any other means. And in the process the hold that these smaller movements have is diminished and one can begin to see or experience larger vistas of light.

Such a passage where one is always moving to larger vistas of light, more easily allows entry into a field of original seeds and it is so that each can begin to enter into conscious communion with truer individuality.

Light's Emergence as Truer Individuality

As discussed previously the Light-Space-Time Emergence equation (3.1.3) being iterative, can be used to model emergence as it proceeds from simpler four-fold to more complex four-fold manifestations. Hence (3.1.3) has already been applied to suggest the emergence of the electromagnetic spectrum, quantum particles in general, bosons as a further instance of a particular kind of quantum particle, atoms, living cells, and basic capacities of the self. Here it will be applied to suggest the emergence of truer individuality. But further, as implied by (3.6.5), the Quantization Effect of Organization on Material-Fabric equation, any process of deeper organization, such as is responsible for the architecture and cohesiveness of truer individuality, has the possibility of altering the material-fabric or fabric of existence so long as the bases involved are driven primarily by a meta-level.

As suggested by (3.1.3), reproduced below for convenience, the architecture and details of truer individuality can be seen to be the result of the application of the Light, Space, and Time matrices as will be elaborated:

$$Emergence_{light-space-time} =$$

$$
\begin{vmatrix}
\begin{vmatrix}
C_\infty:[Pr, Po, K, H] \\
\left(\downarrow R_{C_K} = f(R_{C_\infty})\right) \\
C_K: [S_{Pr}, S_{Po}, S_K, S_H] \\
\left(\downarrow R_{C_N} = f(R_{C_K})\right) \\
C_N: f(S_{Pr} \times S_{Po} \times S_K \times S_H) \\
\left(\downarrow R_{C_U} = f(R_{C_N})\right) \\
C_U: [P,V,M,C]
\end{vmatrix}_{Light}
&
\begin{vmatrix}
M_3 \to System_X \\
(\uparrow F \to I) \\
M_2 \to S_{System_X} \\
(\uparrow Sig \to F) \\
M_1 \to Sig_x \\
(\uparrow > P_x) \\
U \to x_U
\end{vmatrix}_{Space}
\\
\begin{vmatrix}
M_3: -\infty \leq t \leq \infty \\
\downarrow \\
M_2: 0 \geq t > \infty \\
\downarrow \\
M_1: 0 > t > \infty \\
\downarrow \\
U \to \begin{array}{l} t \leq E_{Cell}; TC: M_3 \to U \\ t \sim E_{Human}; TC: U \to M_3 \end{array}
\end{vmatrix}_{Time}
\quad TC \to x_T
&
\langle x_U | x_T \rangle
\end{vmatrix}
$$

Starting with the Light-Matrix, the top left-hand matrix in (3.1.3), the first line from the top, $C_\infty:[Pr, Po, K, H]$, specifies the architecture of truer individuality as will be further discussed in this chapter. Hence, Knowledge-type individuals are an emergence of Light's property of Knowledge, Power-type individuals are an emergence of Light's property of Power, Service-type individuals are an emergence of Light's property of Presence, and Harmony-type individuals are an emergence of Light's property of Harmony. The fundamental architecture of these aspects hence, is an emergence of the properties of Light at ∞.

Line 3 in the Light-Matrix, $C_K: [S_{Pr}, S_{Po}, S_K, S_H]$, elaborates the sets for Presence, Power, Knowledge, and Harmony, each containing multiple elements. For example, as will be explored in the subsequent subsection, various elements derived from the four sets would define the architecture of Harmony-type individuals and could be functions such as 'driven to connect people together', amongst others, hence collectively describing Harmony-type individuals' way of being. Specifically, Line 5, $C_N: f(S_{Pr} \times S_{Po} \times S_K \times S_H)$, suggests that unique

173

seeds are created from a combination of such elements from all four sets, with a particular element leading, that in effect creates the distinctness possible at the level of individuals.

Line 6, $(\downarrow R_{C_U} = f(R_{C_N}))$, specifies quantization between the layer where the seeds are formed, and the physical layer we are familiar with, and as explored in Chapter 3.5 and 3.6, will result in Line 7, $C_U: [P,V,M,C]$, hence changing the material-fabric of existence. The possibilities represented by Lines 1 through 5 hence concretize through the quantization represented by Line 6 to become or further enhance the unique individuality with its subtle physical (related to Presence), vital (related to Power), mental (related to Knowledge), and connection (related to Harmony) aspects now existing in material reality typified by Light moving at c.   Note that just as Line 6 represents a process of quantization relating the layer of reality created by Light traveling at c with the antecedent layers, so too Lines 2 and 4 as previously discussed, also represent quantization of a more subtle kind that ultimately plays a critical part in allowing the material-fabric to express infinite diversity.

Typically it is the process as captured by the Space-Matrix that will determine if Line 6 is activated. Specifically patterns at the untransformed layer, U, will need to be overcome, as specified by the second-line from the bottom of the Space-Matrix: $(\uparrow > P_x)$. But as specified by the bottom-line of the Time-Matrix, reproduced below, it is only with the advent of the human-system that the automaticity of the action of meta-levels is reversed:

$$U \to \begin{array}{l} t \leq E_{Cell}; TC: M_3 \to U \\ t \sim E_{Human}; TC: U \to M_3 \end{array}$$

Hence in the case of truer individuality, which in this emergence is a post-human system, the fact that patterns do need to be overcome means that quantization requires effort to happen.   Given this, it is useful to review Equation 3.6.5, Quantization Effect of Organization on Material-Fabric:

$$Impact\ on\ Materal\ Fabric = \begin{bmatrix} |[L][S][T]TC \to x_T|_{(x_U \mid x_T)} \\ \times \\ (|mod(Z_Q)|_{Y > U}) \\ \ni \\ Z_Q, Z \in U\ (Space, Time, Energy, Gravity) \end{bmatrix}$$

Line 1 in the matrix is simply (3.6.1) the Simplified Light-Space-Time Emergence equation. To understand the quantization that may occur, the organization resulting from the application of Line 1 is multiplied ($\times$) by a modulation (mod) of a fourfold space-time-energy-gravity quantization (Z_Q), so long as Y>U, that is, the operative bases are relatively transformed and hence have an active influence greater than U. The fourfold quantization is specified by ($\ni Z_Q, Z \in U \ (Space, Time, Energy, Gravity))$ where ($\ni$) the quantization being applied (Z) is each (U) of the elements of the set (Space, Time, Energy, Gravity).

But as just summarized in the Time-Matrix in (3.1.3) Y is by definition not greater than U and hence quantization is not automatic. In terms of truer individuality such quantization implies that increasing wholeness can become fully active through specific space, time, energy, and gravity quantization to create an holistic "ecosystem" with its own "truer individuality logic" as it were. The wholeness only then precipitates into the material-fabric and is available to be consciously and unconsciously tapped into.

Precipitation of Individual Types Logic into Material-Fabric

The truer individuality itself is some mathematical function of the four properties of Light - Harmony, Power, Knowledge, and Service, which in turn, are practically infinite sets of qualities related to one of the main properties of Light.

So it may be that one individual is primarily driven to connect people together, itself a variation of the property of Harmony. The individual may further want to do this by deeply understanding what makes these people tick, itself a variation of the property of Knowledge. So the seed or the truer individuality of this person can be thought of as a mathematical function consisting of some element or property from the set of Harmony and some element or quality from the set of Knowledge, combined together with possibly different elements from the same or different sets, all with possibly different weights, but with the first weight of the need to connect people, being the strongest.

Such an individual may have an essential form as expressed in Equation 7.3.1, Harmony-Type Individual:

175

$$Sig_{Harmony-type} = Xa + Yb_{0-n}^{-} \quad where \begin{bmatrix} X \in [S_{System_N}] \\ Y \in [S_{System_{Pr}}, S_{System_P}, S_{System_K}, S_{System_N}] \\ a, b \ are \ integers; a > b \end{bmatrix}$$

Eq 7.3.1: Harmony-Type Individual

But equally, and consistent with the emergent properties of Light there could be essential Service-Type, Power-Type, and Knowledge-Type individuals as well, with infinite variation in the precise nature of the core seed, as captured in Equations 7.3.2-4:

$$Sig_{Service-type} = Xa + Yb_{0-n}^{-} \quad where \begin{bmatrix} X \in [S_{System_{Pr}}] \\ Y \in [S_{System_{Pr}}, S_{System_P}, S_{System_K}, S_{System_N}] \\ a, b \ are \ integers; a > b \end{bmatrix}$$

Eq 7.3.2: Service-Type Individual

$$Sig_{Power-type} = Xa + Yb_{0-n}^{-} \quad where \begin{bmatrix} X \in [S_{System_P}] \\ Y \in [S_{System_{Pr}}, S_{System_P}, S_{System_K}, S_{System_N}] \\ a, b \ are \ integers; a > b \end{bmatrix}$$

Eq 7.3.3: Power-Type Individual

$$Sig_{Knowledge-type} = Xa + Yb_{0-n}^{-} \quad where \begin{bmatrix} X \in [S_{System_K}] \\ Y \in [S_{System_{Pr}}, S_{System_P}, S_{System_K}, S_{System_N}] \\ a, b \ are \ integers; a > b \end{bmatrix}$$

Eq 7.3.4: Knowledge-Type Individual

Note that (7.3.1 - 4) already implies that Lines 1 – 5 in the Light Matrix (3.1.3) have been activated, and that the logic of the truer-individuality-ecosystem will automatically precipitate into the material-fabric through the action of Line 6-7 of (3.1.3).

SECTION 8: COMPUTING COMPLEX ORGANIZATION

Having traced the computations resulting in the emergence of space-time-energy-gravity, the electromagnetic spectrum, matter, life, and even human becoming as manifest in sensation, urges, feelings, and thoughts, and truer individuality, we now turn our attention to study the computation resulting in the emergence of complex organizations.

Chapter 8.1 examines the emergence of mega-organizations. Chapter 8.2 examines the emergence of sustainable global civilization.

Chapter 8.1: Quantum-Level Computation in the Creation of a Stable Mega-Organization Ecosystem

We have traced the emergence of the four properties of Light through the primary fourfold emergence of space-time-energy-gravity, the electromagnetic spectrum, through matter, through life, and even through human becoming as manifest in sensation, urges, feelings, and thoughts, and truer individuality. At each of these levels it seems it is a balanced combination of all four of the properties that creates stability and makes that structure or organism or being sustainable. And this has to be since the four properties occur simultaneously in light and must be impelled to hold to that relationship even when projected from it. Hence it should be possible to deductively arrive at an equation for organizational sustainability given the four-fold emergence of properties of light through subsequent layers of manifestation.

To triangulate, though, we briefly trace the four-fold emergence, reinforce the significance of this and inductively arrive at an equation for sustainability that we would have arrived at deductively anyway.

Starting with the primary emergence of space-time-energy-gravity we see that space is defined as that in which seeds of knowledge are planted – hence, existing as a means to express Light's property of knowledge. Time is defined as the inevitability of the seeds of knowledge maturing into fullness, in spite of any opposition. Hence time is seen as an emergence of Light's property of power. Energy is seen as that which accumulates to create matter. Hence it is an emergence of Light's property of Presence. Gravity is that by which relationship between objects comes into being – hence an emergence of Light's property of harmony.

With the electromagnetic spectrum we see that the spectrum as a whole is defined or architected by the speed of light c, as an emergence of nurturing, the wavelength λ, as an emergence of knowledge, the frequency f, as an emergence of power, and mass-potential as an emergence of presence. These occur simultaneously and constitute the wholeness of the electromagnetic spectrum.

Similarly, at the level of the quantum particle it can be seen that the integrity and functioning of the atom also depends on the integration of all four properties of light. Hence quarks, an emergence of knowledge, leptons, an emergence of power, bosons, an emergence of harmony or nurturing, and the Higgs-boson, an emergence of presence, act together to create the structure and functionality of every single atom.

What this reinforces is that every single thing must have been created through some integration and balance of the four properties of light.

Going up the organizational scale to the next level of complexity, to the level of the atom, the same pattern is found again: every single atom belongs to one of four groups, and in unison these four groups orchestrate the set of combinations of known compounds.

Hence, Alkali Metals and Alkali Earth Metals configured by the s-Group, from the family of system-power, Metals, Metalloid, Non-Metals, Halogens, and Noble Gases configured by the p-Group, from the family of system-knowledge, Transition Metals configured by the d-Group, from the family of system-presence, and Lanthanides and Actinides configured by the f-Group, from the family of system-nurturing, act together to create the complex array of compounds that form the entire material from which the physical and subsequently even the organic world around us is constructed.

Going up the organizational scale to another level of complexity, to the level of the cell, the same pattern is found again: basically every single cell of every single living creature that has ever been studied by humankind also has a similar balance and integration of these four sets of forces acting together.

Hence, proteins, from the family of presence, nucleic acids, from the family of knowledge, polysaccharides, from the family of power, and lipids, from the family of nurturing, work together to create the balanced functioning of every living cell.

The hypothesis hence, is that even at larger scale, in considering the most innovative organization at the level of teams, corporations, or markets, a similar integration and balance of the four sets of properties or forces may yield the best results. Recent developments in sustainability investment models ranging from Socially Responsible Investing (Logue, 2008), the Global Reporting Initiative (Willis, 2003), the Dow Jones Sustainability Index (Hope & Fowler, 2007), the Principles of Responsible Investing (Harvard-Edu, 2014), all reinforce the concept of investment criteria becoming broader to be based not only on economic, but on environmental, social, and governance factors as well. Further evidence suggests that financial returns on such broad-based investment models

179

continually beat financial returns on regular funds such as the S&P 500, for instance (Openshaw, 2015). In their study of major transitions in evolution Smith and Szathmary (Smith & Szathmary, 1995) chronicle the development of life toward increasing complexity as an application of successful collaboration and even co-evolution by which species evolve by changing together as system pressures increase. This hypothesis can be summarized by the following graph:

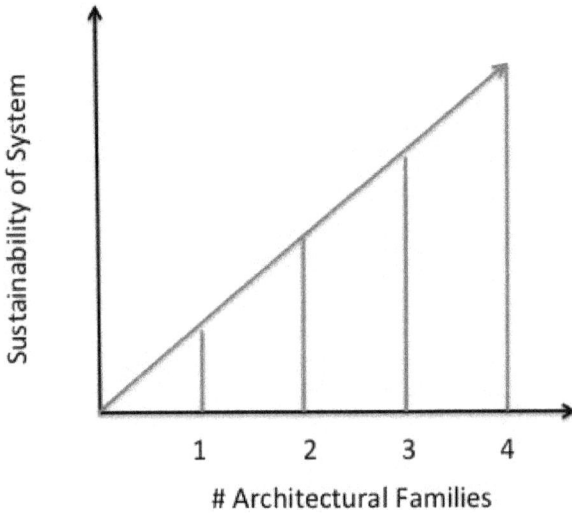

Figure 8.1.1 Sustainability of CAS

An equation for the sustainability of systems, $Sustainability_{Systems}$, where the interaction between the four families of forces is instrumental can be created. Hence, as in Equation 8.1.1:

$$Sustainability_{Systems} \propto Interaction \, (S_{System_{Pr}}, S_{System_P}, S_{System_K}, S_{System_N})$$

Eq 8.1.1: Sustainability of Systems

This notion of the balance of the fourfold properties of Light as fundamental in creating sustainable mega-organization will be borne out in the next subsections of this chapter.

Light's Emergence as Mega-Organization

As discussed previously the Light-Space-Time Emergence equation (3.1.3) being iterative, can be used to model emergence as it proceeds from simpler four-fold to more complex four-fold manifestations. Hence (3.1.3) has already been applied to suggest the emergence of the electromagnetic spectrum, quantum particles in general, bosons as a further instance of a particular kind of quantum particle, atoms, living cells, basic capacities of the self, and truer individuality. Here it will be applied to suggest the emergence of mega-organization. But further, as implied by (3.6.5), the Quantization Effect of Organization on Material-fabric equation, any process of deeper organization, such as is responsible for the architecture and cohesiveness of mega-organizations, has the possibility of altering the material-fabric or fabric of existence so long as the bases involved are driven primarily by a meta-level.

As suggested by (3.1.3), reproduced below for convenience, the architecture and details of the mega-organization can be seen to be the result of the application of the Light, Space, and Time matrices as will be elaborated:

$Emergence_{light-space-time} =$

$$
\begin{vmatrix}
\begin{vmatrix}
C_\infty : [Pr, Po, K, H] \\
(\downarrow R_{C_K} = f(R_{C_\infty})) \\
C_K : [S_{Pr}, S_{Po}, S_K, S_H] \\
(\downarrow R_{C_N} = f(R_{C_K})) \\
C_N : f(S_{Pr} \times S_{Po} \times S_K \times S_H) \\
(\downarrow R_{C_U} = f(R_{C_N})) \\
C_U : [P, V, M, C]
\end{vmatrix}_{Light}
&
\begin{vmatrix}
M_3 \rightarrow System_X \\
(\uparrow F \rightarrow I) \\
M_2 \rightarrow S_{System_X} \\
(\uparrow Sig \rightarrow F) \\
M_1 \rightarrow Sig_x \\
(\uparrow > P_x) \\
U \rightarrow x_U
\end{vmatrix}_{Space}
&
\\
\begin{vmatrix}
M_3 : -\infty \le t \le \infty \\
\downarrow \\
M_2 : 0 \ge t > \infty \\
\downarrow \\
M_1 : 0 > t > \infty \\
\downarrow \\
U \rightarrow \begin{matrix} t \le E_{Cell}; TC: M_3 \rightarrow U \\ t \sim E_{Human}; TC: U \rightarrow M_3 \end{matrix}
\end{vmatrix}_{Time}
&
TC \rightarrow x_T
&
\langle x_U \mid x_T \rangle
\end{vmatrix}
$$

Starting with the Light-Matrix, the top left-hand matrix in (3.1.3), the first line from the top, $C_\infty : [Pr, Po, K, H]$, specifies the architecture of mega-organization as will be discussed shortly. Hence, Academic organizations are an emergence of Light's property of Knowledge, Political organizations are an emergence of Light's property of Power, Social organizations are an emergence of Light's property of Presence, and Commercial organizations can be an emergence of

Light's property of Harmony. The fundamental architecture of these aspects hence, is an emergence of the properties of Light at ∞.

Line 3 in the Light-Matrix, $C_K: [S_{Pr}, S_{Po}, S_K, S_H]$, elaborates the sets for Presence, Power, Knowledge, and Harmony, each containing multiple elements. For example, as will be explored in the following subsection, various elements derived from the four sets would define the architecture of a Silicon Valley type mega-commercial organization and could be functions such as 'commercial immunity', 'power of aesthetics', 'ease of informal meetings', amongst others, hence collectively describing mega-commercial organization's way of being. Specifically, Line 5, $C_N: f(S_{Pr} \times S_{Po} \times S_K \times S_H)$, suggests that unique seeds are created from a combination of such elements from all four sets, with a particular element leading, that in effect creates the distinctness possible at the level of mega-organization.

Line 6, $(\downarrow R_{C_U} = f(R_{C_N}))$, specifies quantization between the layer where the seeds are formed, and the physical layer we are familiar with, and as explored in Chapter 3.5 and 3.6, will result in Line 7, $C_U: [P,V,M,C]$, hence changing the material-fabric of existence. The possibilities represented by Lines 1 through 5 hence concretize through the quantization represented by Line 6 to become or further enhance the mega-organization with its subtle physical (related to Presence), vital (related to Power), mental (related to Knowledge), and connection (related to Harmony) aspects now existing in material reality typified by Light moving at c. Note that just as Line 6 represents a process of quantization relating the layer of reality created by Light traveling at c with the antecedent layers, so too Lines 2 and 4 as previously discussed, also represent quantization of a more subtle kind that ultimately plays a critical part in allowing the material-fabric to express infinite diversity.

Typically it is the process as captured by the Space-Matrix that will determine if Line 6 is activated. Specifically patterns at the untransformed layer, U, will need to be overcome, as specified by the second-line from the bottom of the Space-Matrix: $(\uparrow > P_x)$. But as specified by the bottom-line of the Time-Matrix, reproduced below, it is only with the advent of the human-system that the automaticity of the action of meta-levels is reversed:

$$U \to \begin{array}{l} t \leq E_{Cell}; TC: M_3 \to U \\ t \sim E_{Human}; TC: U \to M_3 \end{array}$$

Hence in the case of mega-organizations, which in this emergence is a post-human system, the fact that patterns do need to be overcome means that quantization requires effort to happen. Given this, it is useful to review Equation 3.6.5, Quantization Effect of Organization on Material-Fabric:

$$Impact\ on\ Materal\ Fabric = \begin{bmatrix} |[L][S][T]TC \rightarrow x_T|_{\langle x_U \,|\, x_T \rangle} \\ \times \\ (|mod(Z_Q)|_{Y > U}) \\ \ni \\ Z_Q, Z \in U\ (\ Space, Time, Energy, Gravity) \end{bmatrix}$$

Line 1 in the matrix is simply (3.6.1) the Simplified Light-Space-Time Emergence equation. To understand the quantization that may occur, the organization resulting from the application of Line 1 is multiplied $(\times)$ by a modulation (mod) of a fourfold space-time-energy-gravity quantization (Z_Q), so long as Y>U, that is, the operative bases are relatively transformed and hence have an active influence greater than U. The fourfold quantization is specified by $(\ni Z_Q, Z \in U\ (Space, Time, Energy, Gravity))$ where $(\ni)$ the quantization being applied (Z) is each (U) of the elements of the set (Space, Time, Energy, Gravity).

But as just summarized in the Time-Matrix in (3.1.3) Y is by definition not greater than U and hence quantization is not automatic. In terms of mega-organizations such quantization implies that increasing wholeness can become fully active through specific space, time, energy, and gravity quantization to create an holistic "ecosystem" with its own "mega-organizations logic" as it were. The wholeness only then precipitates into the material-fabric and is available to be consciously and unconsciously tapped into.

Precipitation of Mega-Organization Logic into Material-Fabric

As suggested by inductively derived (8.1.1), a sustainable organization is the result of two dynamics: maturity along each of the architectural force dimensions, and a robust interaction between all for sets of forces. These dynamics are implicit in (3.1.3) as borne out in the previous subsection.

Consider the example of Silicon Valley.

While Silicon Valley may be primarily a commercial organization, yet we begin to glimpse something of what may become possible when the four emergent

183

forces of light can begin to interact with one another in free fashion. That is, without the motive of doing everything for the sake of generating money only. In the case of Silicon Valley we see that the climate and beauty of the Bay Area, thereby likely representing the impetus of nurturing and harmony, began to attract many talented people into the vicinity. Over time the talent pool became progressively diversified. Educational institutes, such as Stanford University and University of Berkeley cropped up and became centers of cutting-edge research, representing the impetus of knowledge. The Armed Forces were attracted to the area for the same reason, and came with their huge requirements for research for research's sake, and their funds in support of the area, representing the impetus of power. Graduates from the universities started companies that began to in turn support the universities with funds. Talent moved around from company to company like people from department to department. Thus we see that even when individual companies failed, Silicon Valley functioned as a larger organization and was able to retain the talent in the area, was able further, to buffer the shocks to some extent, hence preparing the ground for future waves of innovation. This reality of becoming a container for dynamic experimentation perhaps represents the impetus of service in that the container is serving the dynamic components existing within it.

Pragmatically speaking this dynamic of functioning as a larger mega-organization meant that some level of commercial-immunity began to develop, so that people losing their jobs was not necessarily considered as that stressful an event. There was a freeing, thus, from the purely commercial element. If the various powers of beauty, aesthetics, knowledge, power in the form of money, plus the myriad streams of talent – from engineers, scientists, managers, lawyers etc., did not exist in such close proximity, and further, if the various professionals could not thus support each other through their continued informal meetings, through the urge to create the next wave of innovation, through the urge to pursue progress for the sake of progress, then they would have left for other areas and the phenomenon of Silicon Valley would never have been.

Usually modern day organizations cannot provide these various buffers and opportunities for interaction that Silicon Valley provides, and hence when hit with adversity, more often than not, simply crumble. This is perhaps due to the fact that it is always only one-dimension that drives them, and when the health of that is threatened or falls, the organization resorts to tactics that will ensure that that dimension looks good at any cost, in the bargain sacrificing the development of the other dimensions, and therefore its longer term health.

The four-forces led organization will be something like a community. Having attained freedom from an exaggerated commercial impetus, people will 'live' their 'jobs' because it is what fulfills them. Such a freedom is what will allow the four primary powers to manifest to greater and greater degree within them and their environment. Seeing thus, their ability to become centers of knowledge and wisdom, or mutuality and harmony, or power and leadership and energy, or perfection and service, or some unique combination of these primary forces, so increase, a sense of satisfaction with life will more easily accompany all that they continue to do. Under the freer flow of these powers their uniqueness will be refined and flourish, and correspondingly, so too will the uniqueness of their respective organizations.

An organization may be political and hence be primarily driven by the power of leadership and courage, or it may be social and hence be primarily driven by the power of service and perfection, or it may be commercial and hence be primarily driven by the power of mutuality and harmony, or it may be research-oriented and academic and hence be primarily driven by the powers of knowledge and wisdom, but always each of the other powers will also be behind it, fulfilling and completing its primary urge. Thus the dynamic of unique primary powers being supported by an increasing plethora of secondary powers is represented by the following four organizational equations, (8.1.2 – 5) and illustrates how the four powers of light emerge even in complex human-based collectivities:

$$Sig_{Political} = Xa + Yb_{0-n}^- \quad where \begin{bmatrix} X \in [S_{System_P}] \\ Y \in [S_{System_{Pr}}, S_{System_P}, S_{System_K}, S_{System_N}] \\ a, b\ are\ integers; a > b \end{bmatrix}$$

Eq 8.1.2: Political Organization

$$Sig_{Social} = Xa + Yb_{0-n}^- \quad where \begin{bmatrix} X \in [S_{System_{Pr}}] \\ Y \in [S_{System_{Pr}}, S_{System_P}, S_{System_K}, S_{System_N}] \\ a, b\ are\ integers; a > b \end{bmatrix}$$

Eq 8.1.3: Social Organization

$$Sig_{Commercial} = Xa + Yb_{0-n}^- \quad where \begin{bmatrix} X \in [S_{System_N}] \\ Y \in [S_{System_{Pr}}, S_{System_P}, S_{System_K}, S_{System_N}] \\ a, b\ are\ integers; a > b \end{bmatrix}$$

Eq 8.1.4: Commercial Organization

$$Sig_{Academic} = Xa + Y\overline{b}_{0-n} \quad where \begin{bmatrix} X \in [S_{System_K}] \\ Y \in [S_{System_{Pr}}, S_{System_P}, S_{System_K}, S_{System_N}] \\ a, b \ are \ integers; a > b \end{bmatrix}$$

Eq 8.1.5: Academic Organization

Note that (8.1.2 - 5) already implies that Lines 1 – 5 in the Light Matrix (3.1.3) have been activated, and that the logic of the mega-organization-ecosystem will automatically precipitate into the material-fabric through the action of Line 6-7 of (3.1.3).

Leveraging Mathematical Operators to Promote Organizational and Ecosystem Evolution

Note too that the host of mathematical operators derived deductively in Chapter 2.8, and represented by (2.8.1-17) can also be leveraged to support the emergence of the deeper powers represented by (8.1.2-5) and further enhance the overall ecosystem to operate in a more transformative way. These operators have to be considered in context to the Generalized Equation for Innovation, Equation 2.6.6, derived in Chapter 2.6., or more fully in context of (3.1.3). This equation suggests that any system has implicit in it the urge to transform the untransformed layer, U, by opening to the influence of the meta-layers, M_1, M_2, and M_3. In so doing the very sources of innovation are altered and the visible characteristics of systems are transformed by the action of these sources of innovation. Several sets were suggested that explore these sources of innovation - $S_{System_{Pr}}$, S_{System_P}, S_{System_K}, and S_{System_N}, and the resultant characteristics of systems - $Physical_T$, $Vital_T$, $Mental_T$, and $Integral_T$.

Even the number of "unique" people will continue to increase, become guiding lights so that the community will spontaneously begin to move in holistic directions consistent with the urge of people, of the sub-organizations within the community, of the community itself, and of the larger system of which the community or mega-organization is a part.

Even perhaps, something that may be behind the four archetypal powers, something that may be behind the building of all uniqueness – individual and organizational, more of the fullness of light and love, that is, may be compelled to come forward, seeing how developed its means of expression and action have become.

Chapter 8.2: Quantum-Computation in the Creation of a Sustainable Global Civilization Ecosystem

This chapter will consider the quantum-level computation necessary in the creation of a sustainable global civilization.

Light's Emergence as Sustainable Global Civilization

As discussed in previous chapters the Light-Space-Time Emergence equation (3.1.3) being iterative, can be used to model emergence as it proceeds from simpler four-fold to more complex four-fold manifestations. Hence (3.1.3) has already been applied to suggest the emergence of the electromagnetic spectrum, quantum particles in general, bosons as a further instance of a particular kind of quantum particle, atoms, living cells, basic capacities of the self, truer individuality, and mega-organizations. Here it will be applied to suggest the emergence of sustainable global civilization. But further, as implied by (3.6.5), the Quantization Effect of Organization on Material-Fabric equation, any process of deeper organization, such as is responsible for the architecture and cohesiveness of sustainable global civilization, has the possibility of altering the material-fabric or fabric of existence so long as the bases involved are driven primarily by a meta-level.

As suggested by (3.1.3), reproduced below for convenience, the architecture and details of sustainable global civilization can be seen to be the result of the application of the Light, Space, and Time matrices as will be elaborated:

$Emergence_{light-space-time} =$

$$\left| \begin{array}{c} \left| \begin{array}{c} C_\infty:[Pr, Po, K, H] \\ \left(\downarrow R_{C_K} = f(R_{C_\infty})\right) \\ C_K: [S_{Pr}, S_{Po}, S_K, S_H] \\ \left(\downarrow R_{C_N} = f(R_{C_K})\right) \\ C_N: f(S_{Pr} \times S_{Po} \times S_K \times S_H) \\ \left(\downarrow R_{C_U} = f(R_{C_N})\right) \\ C_U: [P, V, M, C] \end{array} \right|_{Light} \quad \left[\begin{array}{c} M_3 \rightarrow System_X \\ (\uparrow F \rightarrow I) \\ M_2 \rightarrow S_{System_X} \\ (\uparrow Sig \rightarrow F) \\ M_1 \rightarrow Sig_x \\ (\uparrow > P_x) \\ U \rightarrow x_U \end{array} \right]_{Space} \\ \begin{array}{c} M_3 : -\infty \le t \le \infty \\ \downarrow \\ M_2 : 0 \ge t > \infty \\ \downarrow \\ M_1 : 0 > t > \infty \\ \downarrow \\ U \rightarrow \begin{array}{c} t \le E_{Cell}; TC: M_3 \rightarrow U \\ t \sim E_{Human}; TC: U \rightarrow M_3 \end{array} \end{array} \Big|_{Time} \; TC \rightarrow x_T \end{array} \right| \quad \langle x_U \mid x_T \rangle$$

Starting with the Light-Matrix, the top left-hand matrix in (3.1.3), the first line from the top, $C_\infty:[Pr, Po, K, H]$, specifies the architecture of sustainable global civilization as will be further discussed in this chapter. A key to such a civilization is also pointed to by the previously derived equation (8.1.1), reproduced here for convenience:

$$Sustainability_{Systems} \propto Interaction\,(S_{System_{Pr}}, S_{System_P}, S_{System_K}, S_{System_N})$$

This requires maturity by large organized parts of the world, whether nations or regional blocs. The maturity is such that the fourfold properties of Light are adequately expressed. Hence in the example of nations to be illustrated shortly, a country like India is in its deeper essence an emergence of Light's property of Knowledge, a country like Japan is in its deeper essence an emergence of Light's property of Power, a country like Thailand is in its deeper essence an emergence of Light's property of Presence, and a country like UK is in in its deeper essence an emergence of Light's property of Harmony. The fundamental architecture of the combination of such emergences, which as per (8.1.1) is required for sustainability, is hence an emergence of the properties of Light at ∞.

Line 3 in the Light-Matrix, $C_K: [S_{Pr}, S_{Po}, S_K, S_H]$, elaborates the sets for Presence, Power, Knowledge, and Harmony, each containing multiple elements. For example, as will bet explored, various elements derived from the four sets define the architecture of a knowledge-essence country like India and could be functions such as 'exceptional capacity for penetrating behind the surface',

189

'meaningfully synthesizing many streams of development', amongst others, hence collectively describing different aspects of a knowledge-essence country's way of being. Specifically, Line 5, $C_{N:}\, f\left(S_{Pr} \times S_{Po} \times S_K \times S_H\right)$, suggests that unique seeds are created from a combination of such elements from all four sets, with a particular element leading, that in effect creates the distinctness possible at the level of sustainable global civilization.

Line 6, $(\downarrow R_{C_U} = f(R_{C_N}))$, specifies quantization between the layer where the seeds are formed, and the physical layer we are familiar with, and as explored in Chapter 3.5 and 3.6, will result in Line 7, $C_{U:}\, [P,V,M,C]$, hence changing the material-fabric of existence. The possibilities represented by Lines 1 through 5 hence concretize through the quantization represented by Line 6 to enhance sustainable global civilization with subtle physical (related to Presence), vital (related to Power), mental (related to Knowledge), and connection (related to Harmony) aspects now existing in material reality typified by Light moving at c. Note that just as Line 6 represents a process of quantization relating the layer of reality created by Light traveling at c with the antecedent layers, so too Lines 2 and 4 as previously discussed, also represent quantization of a more subtle kind that ultimately plays a critical part in allowing the material-fabric to express infinite diversity.

Typically it is the process as captured by the Space-Matrix that will determine if Line 6 is activated. Specifically patterns at the untransformed layer, U, will need to be overcome, as specified by the second-line from the bottom of the Space-Matrix: $(\uparrow > P_x)$. But as specified by the bottom-line of the Time-Matrix, reproduced below, it is only with the advent of the human-system that the automaticity of the action of meta-levels is reversed:

$$U \to \begin{array}{l} t \le E_{Cell};TC: M_3 \to U \\ t \sim E_{Human};TC: U \to M_3 \end{array}$$

Hence in the case of sustainable global civilization, which in this emergence is a post-human system, the fact that patterns do need to be overcome means that quantization requires effort to happen. Given this, it is useful to review Equation 3.6.5, Quantization Effect of Organization on Material-Fabric:

$$\text{Impact on Materal Fabric} = \begin{bmatrix} |[L][S][T]TC{\rightarrow}x_T|_{\langle x_U | x_T\rangle} \\ \times \\ (|mod(Z_Q)|_{Y>U}) \\ \ni \\ Z_Q, Z \in U \ (Space, Time, Energy, Gravity) \end{bmatrix}$$

Line 1 in the matrix is simply (3.6.1) the Simplified Light-Space-Time Emergence equation. To understand the quantization that may occur, the organization resulting from the application of Line 1 is multiplied $(\times)$ by a modulation (mod) of a fourfold space-time-energy-gravity quantization (Z_Q), so long as Y>U, that is, the operative bases are relatively transformed and hence have an active influence greater than U. The fourfold quantization is specified by $(\ni Z_Q, Z \in U \ (Space, Time, Energy, Gravity))$ where $(\ni)$ the quantization being applied (Z) is each (U) of the elements of the set (Space, Time, Energy, Gravity).

But as just summarized in the Time-Matrix in (3.1.3) Y is by definition not greater than U and hence quantization is not automatic. In terms of sustainable global civilization such quantization implies that increasing wholeness can become fully active through specific space, time, energy, and gravity quantization to create an holistic "ecosystem" with its own "sustainable global civilization logic" as it were. The wholeness only then precipitates into the material-fabric and is available to be consciously and unconsciously tapped into.

Precipitation of Global Sustainable Civilization Logic into Material-Fabric

A brief look at history will reinforce the idea that it is typically maturity along multiple and distinct dimensions, represented by the four properties of light, and their rich interaction that allows civilizations to endure.

Thus, those civilizations that have endured typically have a balance of all four families (Sri Aurobindo, 1971). Civilizations that have become extinct typically have had a focus on few drivers of innovation. Jared Diamond proposes five interconnected causes of collapse that may reinforce each other: non-sustainable exploitation of resources, climate changes, diminishing support from friendly societies, hostile neighbors, and inappropriate attitudes for change (Diamond, 2005). But these five sources may also be thought of as symptoms that arise due to the failure to adopt the catholicity of the sources of innovation emanating from each of the four sets or families. Further, the historian Toynbee suggested that societies decay because of their over-reliance on structures that helped them solve old problems (Toynbee, 1961). It can be interpreted that being thus biased they are unable to adopt the catholicity of the sources of innovation emanating from each of the four sets of families.

Approaching Civilization from a big-picture, global basis though, the sustainability of humankind will be ensured by a balance of development amongst the four sets of forces. This means that countries and global regions must be unique and in such a way that their primary emergence is distributed amongst all four sets of forces. Further, and based on this uniqueness, there must be an open and healthy interaction amongst these centers of uniqueness.

Hence, India, Japan, Thailand, and UK will be considered as representative examples of the four distinct and emergent properties that must be balanced in the whole.

Historically at least, India, for example, in its essence, may be thought of as having an exceptional capacity for penetrating behind the surface, and further of meaningfully synthesizing many streams of development. Its primary power may thus be from the family of knowledge, with a strong secondary driver being its ability to create living harmonies. The equation for India will then likely be represented by Equation 8.2.1:

$$Sig_{India} = Xa + Yb_{0-n}^{-} \quad where \begin{bmatrix} X \in [S_{System_K}] \\ Y \in [S_{System_{Pr}}, S_{System_P}, S_{System_K}, S_{System_N}] \\ a, b \ are \ integers; a > b \end{bmatrix}$$

Eq 8.2.1: India (Knowledge Family)

Japan, in its essence, may be thought of as having a strong and noble warrior nature, along with a refined sense of aesthetics, amongst other qualities. Its equation, Equation 8.2.2 would be of the form:

$$Sig_{Japon} = Xa + Yb_{0-n}^{-} \quad where \begin{bmatrix} X \in [S_{System_P}] \\ Y \in [S_{System_{Pr}}, S_{System_P}, S_{System_K}, S_{System_N}] \\ a, b \ are \ integers; a > b \end{bmatrix}$$

Eq 8.2.2: Japan (Power Family)

UK, in its essence, may be thought of as having a strong ability to create practical, materialistic harmonies, resulting in such things as working parliaments and advanced democracy, for example. Its equation, Equation 8.2.3, would be of the form:

$$Sig_{UK} = Xa + Yb_{0-n}^{-} \ where \begin{bmatrix} X \in [S_{System_N}] \\ Y \in [S_{System_{Pr}}, S_{System_P}, S_{System_K}, S_{System_N}] \\ a, b \ are \ integers; a > b \end{bmatrix}$$

Eq 8.2.3: UK (Nurturing Family)

Thailand, in its essence, may be characterized by an exceptional sense of hospitality and sweet service, with an attention to detail in the practical arrangement of things. Its equation, Equation 8.2.4, would be of the form:

$$Sig_{Thailand} = Xa + Yb_{0-n}^{-} \ where \begin{bmatrix} X \in [S_{System_N}] \\ Y \in [S_{System_{Pr}}, S_{System_P}, S_{System_K}, S_{System_N}] \\ a, b \ are \ integers; a > b \end{bmatrix}$$

Eq 8.2.4: Thailand (Service Family)

Note that (8.2.1 - 4) already implies that Lines 1 – 5 in the Light Matrix (3.1.3) have been activated, and that the logic of the sustainable-global-civilization-ecosystem will automatically precipitate into the material-fabric through the action of Line 6-7 of (3.1.3).

Similarly every nation on earth will have a uniqueness that can be represented by equations belonging to one of the four families.

As discussed in Chapter 2.5 on Emergence of Uniqueness, for uniqueness to emerge and mature is a process. The process is represented by Equation 2.5.1 reproduced here for convenience:

$$Sig_E = X \begin{vmatrix} C{:}Sig * mod \left(\int = 1 \right) \\ F{:}Sig \ mod \ (c) \\ I{:} Sig \ mod \left(\int G, \bar{e}, \pi \right) \\ M{:}Sig * mod \ (G) \\ V{:}Sig * mod \ (e) \\ P{:}Sig * mod \ (\pi) \end{vmatrix}$$

Equations 8.2.1 – 4 represent example of the essence of uniqueness. For these to become living practicalities requires work at many different levels within a nation. If a nation has not really done the work to animate itself with its uniqueness then it may exist in the physical or P-state, or vital of V-state, or mental or M-state, which by definition lack sufficient maturity, and interaction with other nations is then going to be compromised. Moving to the integral or I-state, will allow a nation to at least not be locked in to a point of view. A consistent I-state practiced by each nation then is the minimum requirement for a sustainable global civilization. But a consistent I-state also means that restricting patterns are always being broken and that the quantization effect of organization on the material-fabric is actively in play. In other words, separations of light are continually uniting with the larger continent of Light and changing the very nature of possibility on larger and larger scale. The next sub-section examines such a quantization effect in more detail.

Leveraging Mathematical Operators to Promote Organizational and Global Ecosystem Evolution

Note too that the host of mathematical operators derived deductively in Chapter 2.8, and represented by (2.8.1-17) can also be leveraged to support the emergence of the deeper powers represented by (8.2.1-4) and further enhance the overall global ecosystem to operate in a more transformative way. These operators have to be considered in context to the Generalized Equation for Innovation, Equation 2.6.6, derived in Chapter 2.6. and (3.1.3) These equations suggest that any system has implicit in it the urge to transform the untransformed layer, U, by opening to the influence of the meta-layers, M_1, M_2, and M_3. In so doing the very sources of innovation are altered and the visible characteristics of systems are transformed by the action of these sources of innovation. Several sets were suggested that explore these sources of innovation - $S_{System_{Pr}}$, S_{System_P}, S_{System_K}, and S_{System_N}, and the resultant characteristics of systems - $Physical_T$, $Vital_T$, $Mental_T$, and $Integral_T$.

Section 9: Rethinking Quantum Computation

Having elaborated on alternative ways to think of superposition, entanglement, and the quantum, and proposed a cohesive light-based mathematics by which quantum computation continues to compute all emergence of Light, this Section will begin to explore alternative ways in which the field of quantum computation may be envisioned.

Chapter 9.1 summarizes four-fold complexification as possibility in Light continues to emerge, also suggesting an alternative stratum for quantum-computation - that of the quantum-level material-fabric, introduced while discussing quantization of space, time, energy, and gravity in Chapter 3.5.

Chapter 9.2 will explore a philosophy of computation by contrasting the digital, qubit, and material-fabric stratums for computation, and some implications.

Chapter 9.3 will explore some alternative paradigms for computation by focusing on the implicit characteristics of the material-fabric stratum.

Sections 4 through 8 illustrate the common Equation of Light-Space-Time Emergence (3.1.3) that can model reality from the Big Bang to the present, and from the micro to the macro. This modeling involves multiple layers of Light, connected as it were through the device of quanta. As such, any and every emergence is seen to involve the quantum-level, and precisely, through a process of continual quantum computation involving quantization. The quantum computation arbitrates, through a process of qualified determinism and considering inputs from multiple layers of reality created by Light traveling at different speeds, the effective output that will become part of practical reality.

It is not a random process from infinite superposed possibilities that exist at the quantum-level as supposed by the Copenhagen Interpretation, and as assumed as the foundation of the infinite processing capability of quantum-objects by contemporary pioneers of quantum computing, but a concord of possibility stacked in logical arrangement of superposition and entanglement, that determines output.   And as illustrated by the prior Sections 4 through 8, such quantum computation does not just happen at the quantum level, but can happen at any and every level of granularity through a process that also simultaneously involves the quantum-levels.

The urge to break all boundaries, which is no doubt a good thing, has perhaps given rise to the current model of quantum computation where superposed quantum objects forming the edifice of an extraordinary qubit-based stratum has been invested with infinite processing power.   But as this chapter will summarize, the effective process of quantum computation that has given rise to all things is a result of multiple layers of Light, and as such a physical-only quantum object, such as an electron or quark, likely cannot house in itself the possibility that is being proposed by today's quantum computer pioneers. It is a different stratum – in previous chapters referred to as the "material-fabric" and introduced in Chapter 3.5 in the discussion on space, time, energy, and gravity quantization – that is enhanced by the computation of (3.1.3) that must be at the core of all quantum computation.

This being the case, it is suggested that the field of quantum computation be rethought. For now, the "material-fabric" is summarized through the process of fourfold complexification reviewed in this chapter.

Space-Time-Energy-Gravity Four-Foldness

In looking at the nature of the fourfold order it is apparent that fourfold-wholeness is more tightly related to a smaller set of fundamental attributes, as will be elaborated, at earlier stages of emergence.

Hence in thinking through the first apparent space-time-energy-gravity emergence, these may be summarized by the equations, Equations 9.1.1 through 9.1.4. In each case there is a dependence on the single fundamental attribute of 'seeds'.

Equation 9.1.1, System-Knowledge (Space) at Space-Time-Energy-Gravity Level, summarizes how knowledge is related to 'infinity of seeds' with each seed representing, as it were, a different arrangement of knowledge:

$$System_{Knowledge} \propto [f(infinity\ of\ seeds)]$$

Eq 9.1.1: System-Knowledge (Space) at Space-Time-Energy-Gravity Level

Note that the infinity of seeds requires 'space' to be expressed.

Equation 9.1.2, System-Power (Time) at Space-Time-Energy-Gravity Level, summarizes how power is related to 'maturation of seeds', since anything that gets in the way of the maturation will generally be overcome:

$$System_{Power} \propto [f(maturation\ of\ seeds)]$$

Eq 9.1.2: System-Power (Time) at Space-Time-Energy-Gravity Level

Note that maturation of seeds is felt in 'time'.

Equation 9.1.3, System-Presence (Energy/Matter) at Space-Time-Energy-Gravity Level, summarizes how presence is related to 'materialization of seeds', since the accumulation of energy will express itself in material form:

$$System_{Presence} \propto [f(materialization\ of\ seeds)]$$

Eq 9.1.3: System-Presence (Energy/Matter) at Space-Time-Energy-Gravity Level

Equation 9.1.4, System-Harmony (Gravity) at Space-Time-Energy-Gravity Level, summarizes how harmony is related to 'relationship of seeds', since seeds will generally have a relationship to one another and likely exist in collectivities:

$System_{Harmony} \propto [f(relationship\ of\ seeds)]$

Eq 9.1.4: System-Harmony (Gravity) at Space-Time-Energy-Gravity Level

Note that such relationship may be expressed as 'gravity'.

The creation of space-time-energy-gravity is a fundamental device that facilitates the process of quantization thereby allowing the accumulated material-fabric to change. It is this material-fabric that is the stratum for quantum-level computation.

Electromagnetic (EM) Spectrum Four-Foldness

In the case of the EM spectrum the number of basic attributes to express four-foldness is two as will be observed.

Equation 9.1.5, System-Harmony at Electromagnetic Level, is related to the speed of light, c at U:

$System_{Harmony} \propto c_U$

Eq 9.1.5: System-Harmony at Electromagnetic Level

Equation 9.1.6, System-Power at Electromagnetic Level, is related to the frequency, v, of the EM spectrum:

$System_{Power} \propto hv$

Eq 9.1.6: System-Power at Electromagnetic Level

Equation 9.1.7, System-Knowledge at Electromagnetic Level, is related to the wavelength, λ, of the EM spectrum:

$System_{Knowledge} \propto [f(\lambda)]$

Eq 9.1.7: System-Knowledge at Electromagnetic Level

And since, $E = mc^2$ or $m = \dfrac{E}{c^2}$, and substituting hv for E, yields Equation 9.1.8, System-Presence at Electromagnetic Level:

$$System_{Presence} \propto \frac{hv}{c^2}$$

Eq 9.1.8: System-Presence at Electromagnetic Level

But, $c \propto \frac{1}{h}$

Hence, the general relationships of the four-fold wholeness may be understood by knowing just two variables, C, and either, λ or v.

The emergence of the EM Spectrum adds another fold as it were to the quantum-level material-fabric stratum that can progressively be seen as a complexification of four-foldness.

Quantum Particle Four-Foldness

As per the model of early cosmic development presented by the Particle Data Group (Particle Data Group, 2015) at time, t = 10^{-10} s, the four-fold fullness begins to express itself in a series of particles. Quarks, leptons, bosons, which also may imply the presence of the Higgs-boson that gives quarks their mass are observed. But as discussed quarks are a precipitation of Light's property of Knowledge, leptons of Power, bosons of Harmony, and the Higgs-boson of Presence. Hence it is again observed that the four-fold fullness has expressed itself at a different order of complexity.

Einstein has suggested that particles are an excitation of the underlying field. The EM spectrum gives insight into a field that propagates at c at different frequencies. But there are other fields too. The Higgs Field has been suggested as being one of these (Jepsen, 2013). The Standard Model suggests that quarks, leptons, bosons, and the Higgs-boson are different fundamental particles, which implies that the fourfold wholeness has now a more complex basis of its operation, since as compared with the fourfold wholeness of the EM spectrum the number of independent bases appears to have increased as summarized in Figure 9.1.1, Complexification of Material-Fabric in Terms of Levels of Implicit Wholeness, to appear below.

Summarizing, as in Equations 9.1.9 through 9.1.12:

$$System_{Harmony} \propto f(bosons)$$

Eq 9.1.9: System-Harmony at Particle Level

$System_{Power} \propto f(leptons)$

Eq 9.1.10: System-Power at Particle Level

$System_{Knowledge} \propto f(quarks)$

Eq 9.1.11: System-Knowledge at Particle Level

$System_{Presence} \propto f(Higgs\_boson)$

Eq 9.1.12: System-Presence at Particle Level

Hence the quantum-particle ecosystem logic is also now added to the quantum-level material-fabric stratum, which as suggested in previous chapters in Sections 5 through 8, progressively continues to complexify along the lines set up by the underlying fourfold organizational principles as dictated by Light's properties.

Periodic Table/Atoms Four-Foldness

As discussed in Chapter 6.3 the Periodic Table itself is also an expression of the four-fold wholeness. Hence, d-Group elements are an instance of Light's property of Presence, p-Group elements are an instance of Light's property of Knowledge, s-Group elements are an instance of Light's property of Power, and f-Group elements are an instance of Light's property of Harmony or Nurturing. As per the Particle Data Group model (Particle Data Group, 2015) this emergence likely began at $t = 3 \times 10^5$ years when the atom is suggested to have emerged, and continued at least to $t = 10^9$ years with the emergence of stars which it is already known are the furnaces in which heavier atoms were created.

Leveraging the fundamentally different groups yields Equations 9.1.13 through 9.1.16:

$System_{Harmony} \propto f(f\_group)$

Eq 9.1.13: System-Harmony at Atomic Level

$System_{Power} \propto f(s\_group)$

Eq 9.1.14: System-Power at Atomic Level

$$System_{Knowledge} \propto f(p\_group)$$

Eq 9.1.15: System-Knowledge at Atomic Level

$$System_{Presence} \propto f(d\_group)$$

Eq 9.1.16: System Presence at Atomic Level

Atoms therefore have at least four four-fold wholenesses implicit in them. The material-fabric from which atoms, along with all previous and to–be-discussed matter and general phenomena arise, therefore further complexifies through the precipitation, as it were, of this additional atom-ecosystem logic.

Living Cell Four-Foldness

At time, $t = 13.8 \times 10^9$ years, there is a fifth clear expression of the same four-fold order as the bases of an even more complex organization, that of cellular life and all that is founded on it. This can be summarized as in Equations 9.1.17 through 9.1.20:

$$System_{Harmony} \propto f(lipids)$$

Eq 9.1.17: System-Harmony at Cellular Level

$$System_{Power} \propto f(polysaccharides)$$

Eq 9.1.18: System-Power at Cellular Level

$$System_{Knowledge} \propto f(nucleic\ acids)$$

Eq 9.1.19: System-Knowledge at Cellular Level

$$System_{Presence} \propto f(proteins)$$

Eq 9.1.20: System-Presence at Cellular Level

Cellular life therefore seems to have at least five four-fold-wholenesses implicit in it to therefore further complexify the architecture of matter and the quantum-level material-fabric from which it arises.

In the mathematical model elaborated in this book, due to Space-Time-Energy-Gravity (S-T-E-G) quantization the reality of matter is always changing. Such quantization is the result of quantum-certainty as discussed in Chapter 3.5, which itself is the result of a cohesive want or will. Since all is Light, as also discussed, Light therefore can change Light by Light. Deductively therefore, a changing matter and material-fabric are inevitable.

Further, as S-T-E-G quantization proposes matter and the underlying quantum-level material-fabric will be changed by seemingly non-physical phenomena. Hence, as new function-elements from the underlying four architectural sets express themselves or are created, matter itself will go through a change. Fundamental capacities of self, hence, will alter the material-fabric or the reality of matter.

Up to the level of living cells there are at least five four-fold-wholenesses implicit in it. As will be proposed here and in subsequent sub-sections in this chapter the number of these implicit four-fold-wholenesses will continue to increase.

Equations 9.1.21 through 9.1.24 summarizes a sixth clear expression of four-foldness that will change the structure of the material-fabric and matter.

$$System_{Harmony} \propto f(Emotions, Feelings)$$

Eq 9.1.21: System-Harmony at Level of Capacities of Self

$$System_{Power} \propto f(Urges, Wills, Desires)$$

Eq 9.1.22: System-Power at Level of Capacities of Self

$$System_{Knowledge} \propto f(Thoughts)$$

Eq 9.1.23: System-Knowledge at Level of Capacities of Self

$$System_{Presence} \propto f(Sensations)$$

Eq 9.1.24: System-Presence at Level of Capacities of Self

Truer Individuality Four-Foldness

Similarly as truer individuality expresses itself S-T-E-G quantization initiated by quantum-certainty will further complexify the bases of matter.

Equations 9.1.25 through 9.1.28 therefore summarize a seventh clear expression of four-foldness that will further complexify the structure of the material-fabric and change the structure of matter.

$$System_{Harmony} \propto f(Harmony - type - individuals)$$

Eq 9.1.25: System-Harmony at Level of Truer Individuality

$$System_{Power} \propto f(Power - type - individuals)$$

Eq 9.1.26: System-Power at Level of Truer Individuality

$$System_{Knowledge} \propto f(Knowledge - type - individuals)$$

Eq 9.1.27: System-Knowledge at Level of Truer Individuality

$$System_{Presence} \propto f(Service - type - individuals)$$

Eq 9.1.28: System-Presence at Level of Truer Individuality

Sustainable Global Civilization Four-Foldness

Similarly as more complex organizational levels, such as a sustainable global civilization expresses itself, S-T-E-G quantization initiated by quantum-certainty will further complexify the quantum-level material-fabric ecosystem logic and the bases of matter. Recall this now brings into matter the capacity represented by creating a sustainable global civilization into its fold as it were to be used consciously or sub-consciously by matter-based entities going forward.

Equations 9.1.29 through 9.1.32 therefore summarize an eighth clear expression of four-foldness that will change the underlying material-fabric and the structure of matter. Note that these expressions are indicative only and there are other missed levels that would need to be expressed for completeness.

$$System_{Harmony} \propto f(Harmony - type - region)$$

Eq 9.1.29: System-Harmony at Level of Nations or Blocs

$$System_{Power} \propto f(Power - type - region)$$

Eq 9.1.30: System-Power at Level of Nations or Blocs

$$System_{Knowledge} \propto f(Knowledge - type - region)$$

Eq 9.1.31: System-Knowledge at Level of Nations or Blocs

$$System_{Presence} \propto f(Service - type - region)$$

Eq 9.1.32: System-Presence at Level of Nations or Blocs

Summary of Complexification of Quantum-Level Material-Fabric

This scheme of complexification is summarized by Figure 9.1.1:

Level	Independent "Variables"	Minimum Levels of Implicit Wholeness	Minimum Number of Levels
Space-Time-Gravity-Energy (S-T-G-E)	Seeds	• S-T-G-E	1
EM Spectrum	c, ν or λ	• S-T-G-E • EM	2
Particle Level	Quarks, Leptons, Bosons, Higgs-Boson	• S-T-G-E • EM • Particle	3
Atomic Level	s-Group, p-Group, d-Group, f-Group	• S-T-G-E • EM • Particle • Atom	4
Cellular Level	Proteins, Nucleic Acids, Lipids, Polysaccharides	• S-T-G-E • EM • Particle • Atom • Cell	5
Level of Capacities of Self	Sensations, Wills, Emotions, Thoughts	• S-T-G-E • EM • Particle • Atom • Cell • Capacities of Self	6

Truer Individuality Level	Service-type-individual, Knowledge-type-individual, Harmony-type-individual, Power-type-individual	<ul><li>S-T-G-E</li><li>EM</li><li>Particle</li><li>Atom</li><li>Cell</li><li>Capacities of Self</li><li>Truer Individuality</li></ul>	7
Sustainable Global Civilization Level	Service-type-region, Knowledge-type-region, Harmony-type-region, Power-type-region	<ul><li>S-T-G-E</li><li>EM</li><li>Particle</li><li>Atom</li><li>Cell</li><li>Capacities of Self</li><li>Truer Individuality</li><li>Sustainable Global Civilization</li></ul>	8

Figure 9.1.1 Complexification of Quantum-Level Material-Fabric in Terms of Levels of Implicit Wholeness

In reference to the four sets at M_2 (as discussed in Chapter 2.3 on Architectural Forces) it may also be that as the basis of matter complexifies as it journeys through the EM spectrum and other potential fields, the elementary particles, the atom, cellular life, complex individual and further organizational development, as summarized in Figure 9.1.1, the sets which have been positioned to each contain infinite elements, concretely manifest more of their function-elements.

This suggestion has been the basis of the equations derived in previous chapters. The complexity of the material-fabric may therefore be related to the number of manifested function-elements of the set of four sets. This idea is consistent with the notion of the "adjacent possible" suggested by Kaufmann (Kaufmann, 2003) in which innovation is positioned as a recombination of existing parts to create new value – or of existing sets to combine parts of themselves to create new elements based on new circumstance. If MS signifies manifested-set, so that the cardinality or number of elements in the combined set is the union of the four manifested-sets, this may be summarized by the following equation, Equation 9.1.33:

$$Material - Fabric_{Complexity} \propto \left| MS_{System_{Pr}} \cup MS_{System_P} \cup MS_{System_K} \cup MS_{System_N} \right|$$

Eq 9.1.33: Complexity of Material-Fabric in Terms of Manifested Set

The quantum-level material-fabric, it will be observed is modeled as a series of four-fold forces, not unlike perhaps, the structure of DNA composed essentially of the fourfold C, G, A, T nucleobases. The material-fabric, hence appears to have two qualities to it. First, as expressed before and as concretized by (3.1.3) it is the effective medium in which quantization takes place. Second, it allows a vast amount of quantization so that in a sense chains of ecosystem-logic can exist.

While it has not been explored in this book, it is possible that the very scheme of DNA animating every living cell, is in fact a concretization of this kind of scheme happening at the quantum-level material-fabric.

Aspects of these themes will be taken up in more detail in the next chapter.

In arriving at computational possibilities, the stratum or reality that is the basis in which the computation will takes place has to be considered. We know that the quantum realm appears different from the day-to-day physical realm, and it stands to reason that therefore the computational possibilities will differ depending on what the characteristics of that realm are. But there is more. The physical realm we are familiar with and on which our every-day laws are based have a more or less familiar stability to them. But the observable physical realm is itself built on the unfamiliar quantum realm. We therefore cannot simply extrapolate the known physical laws to the unknown quantum realms, nor expect that the very basis and object of computation should be the same across these realms.

In this chapter the familiar basis of digital computing will be reviewed. This basis builds off a digital reality that has been the logical construct of computer scientists and is based on established physical phenomena. In addition, the basis of the current efforts in qubit-based computing will briefly be reviewed. These efforts are based on a commonly accepted interpretation of superposition and entanglement that as has been discussed in this book is arguably incomplete.

Finally a quantum-level material-fabric based on a unique interpretation of quanta, of light, of superposition, and of entanglement, and as summarized in Chapter 9.1, is proposed as the context for a different kind of quantum computation.

Digital Computing

In digital computing, an electrical-current through a bit will determine if it takes on a value of 0 or 1. Every bit can have one of two values, and this becomes part of the essential foundation of the binary logic that forms the backbone of common computing devices. Additional logic-gates are created that allow simple mathematical manipulation to be performed on the streams of binary data, thereby further adding to essential logical foundation of digital computing. Networks of gates and binary states are then activated in different ways by computer-programs and cause streams of binary data to flow in myriad ways as dictated by the logic to compute the precise task required by a computer-program in general.

The stratum of digital computing can hence be thought of as these networks of binary-based devices.

By contrast, the contemporary quantum computer pioneer suggests that the stratum of quantum computing is qubit-based, so that rather than being in an either-or binary-state, as in digital computing, a qubit can also be in both states simultaneously. Such simultaneity is suggested as being the outcome of the phenomenon of superposition, as proposed by the Copenhagen Interpretation of Quantum Mechanics. Taking its cue from the dual-slit experiment quantum-objects can supposedly be in infinite states until measured, at which point they collapse into an observable state. The proposition of being in infinite states simultaneously is what is supposed to afford quantum computing with its ability of infinite processing power. But the interference patterns perceived in the dual-slit experiment might also be explained by considering light as having multiple simultaneous realities, with the 'wave' reality guiding the 'particle' reality to the regions of constructive interference as proposed by the De Broglie-Bohm Pilot-Wave Interpretation of Quantum Mechanics, which is also consistent, or perhaps a subset, of the Light-Based Interpretation of Quantum Mechanics as suggested in this book.

Qubit-based quantum computer pioneers will perhaps refer to the validity of all quantum phenomena, citing one renowned phenomenon, the photoelectric effect, for which Albert Einstein received the Nobel Prize. This effect explains why it is not intensity but frequency of light that can dislodge electrons from atoms and sheds insight into Planck's mathematical creation of h, Planck's Constant, and the subsequent insight into quanta. But the validity of one such phenomenon should not give free reign to other quantum phenomena such as superposition and entanglement, without further study and clear proof. Further, it is in our best interest to study and propose different ways in which superposition and entanglement may work, as has been done in this book, before pronouncing a perception as synonymous with reality, and before pronouncing that because of a perceived way in which superposition may work, quantum computers can have infinite processing power.

This notion of building edifice upon edifice is somewhat reminiscent of observations made by Joseph Weizenbaum, the MIT computer scientist, in his book Computer Power and Human Reason in writing about his experiments with ELIZA, a natural language processor he had developed (Weizenbaum, 1976). He states: "This reaction to ELIZA showed me more vividly than anything I had seen hitherto the enormously exaggerated attributions an even well-educated audience is capable of making, even strive to make, to a technology it does not understand."

It is a chimera why Jerome Cordoba, a sixteenth century figure, even is he was a polymath, should be considered as the unacknowledged discoverer of the mathematical foundations of quantum physics (Brooks, 2017). Quantum physics is a twentieth-century development. What does it tell us about the current-state of the field if we have to back a few centuries when there was no conception of quantum physics and when science and math were just beginning a centuries long journey to get to their current levels of relative maturity, to explain all the development of the centuries long journey? Likely that our current conceptions may be misplaced and need to be carefully rethought.

Quantum-Level Material-Fabric Computation

As such, quantum phenomenon such as superposition and entanglement can be conceived of and modeled differently as has been modeled in the Light-Space-Time Emergence Equation (3.1.3). In such a model the stratum of quantum computation then, is not a field of superposed quantum-objects that collapse through the supposed-reality of randomness and probability, but a quantum-level material-fabric that in fact is governed by a scheme such as qualified determinism involving dynamics of multiple layers of light as modeled by (3.1.3).

The material-fabric is such that it is the interface of multiple layers of light and has practically infinite capacity for adaptation and evolution. Further, the myriad fourfold logic-ecosystems, some of which have been described in this book, can exist in it so that matter and all other emergences of Light can consciously and unconsciously tap into them, and even alter them by the process suggested in (3.1.3) to further evolve their own functioning.

The material-fabric may be thought of as housing a Light-based "DNA", perhaps not unlike the DNA that exists in every living-cell, which may even be a precipitation of this essential scheme in more material terms. Elaborating, the two matrices in (3.1.3) – the Light-Matrix and the Space-Matrix are intimately connected. They can in fact be thought of intertwined in a manner such that the Light-Matrix creates realities with specific dynamics based on the speed of light in a descending movement as light slows down from infinity, while the Space-Matrix suggests lower-level dynamics that access higher-level dynamics as conditions are met, therefore describing an ascending movement. But further, the nexus between one layer and the next layer in the descending series in the Light-Matrix is arbitrated by quanta, that in effect are nodes in which implicit properties in Light become more and more material in the descending

209

progression. Hence the intertwining levels are much like the tightly-related "double-helix" as exists in DNA. But further, the dynamics of each level emanates form the underlying fourfold properties just as DNA is comprised of four nucleotides. Finally, the myriad ecosystems that, in this model, precipitate into the material-fabric, suggests too that there is an unending chain of logic that cues dynamics at the material level.

This material-fabric and the logic of the Light-Space-Time Emergence equation then suggests a different stratum for quantum-computation that may give rise to an alternative genre of quantum computers than suggested by contemporary qubit-based quantum computer pioneers.

Sections 4 through 8 in this book traced a journey in which quantum computation has been integral in evolving our universe from the Big Bang to the present time, and from the micro to the macro. This journey suggests a way in which quantum computation, arbitrating between multiple layers of Light as it were, creates logic-ecosystems in the material-fabric that subsequently influences outcomes in material reality. Beyond natural systems that already may use such an approach to computation – such as systems of quantum particle, systems of atoms and molecules, systems of living cells, what is suggested is a new human-quantum genre of computers in which it is a joint effort that determines computation.

Hence, in human-quantum computation the mathematical operators highlighted in Chapter 3.7 may be leveraged so that human attributes of will, imagination, thought, love, can work in unison with computational devices to influence some of the meta-U-level dynamics that will alter the material-fabric.

As reviewed in some detail in this book the universe could very well have been computed into existence. The computation is the result of iteratively applying the Light-Space-Time Emergence equation (3.1.3). Quantization is implicit in this equation and in fact is the operative basis by which the quantum-level material-fabric goes through change. The change itself is a precipitation of additional logic-ecosystems that cue the way different systems will operate at the material level.

As explored in the last chapter these ecosystems can perhaps be thought of as being concatenated to each other in a chain-like manner, and when considered as being the result of the tightly linked Light-Matrix and Space-Matrix, seems to specify a subtle "DNA" in the quantum-level material-fabric itself, especially given the quaternary-bases of all ecosystems. The analogy to DNA hence derives from the quaternary-bases, the tightly linked light and space matrices, and the generated ecosystems that perhaps concatenate in logic-segments.

It is therefore not surprising to learn in Stanford Anthropologist's Narby's book on DNA and the origins of knowledge (Narby, 1998), that across many histories and geographies shamans have perceived detailed representations of DNA in visions. Narby suggests that these seers are able to do this through entering directly into the molecular domain. My contention formed from the mathematical model in this book though, is that such seers have perhaps been able to enter into the material-fabric, that stands at the gateway between layers of reality created by light traveling at different speeds, to experience the subtle-DNA even before it may further precipitate into the material realm.

As a result the base "logic" of the electromagnetic spectrum, of quantum particles, of atoms, of molecules, of living cells and by extrapolation of all living things, of fundamental human capacities, of individuality, of progressively more complex organization, mega-organization, and of a sustainable global civilization seems to be embedded in the material-fabric. Adaptation itself involves an interrelation between multiple levels and as just summarized, an effective process of quantization, followed by alteration of the quantum-level material-fabric itself.

Against this possibly complex computational reality happening in real-time at the quantum-levels, is our belief that the fundamental layer of existence on which all other existence is based, is just the physical quantum-level without any antecedent meta-levels informing it.

Hence we may try to figure or reduce any computational complexity in terms of the measurable quantum numbers – principal, azimuthal, magnetic, spin – or to observable changes in them to explain all the complexity that the Light-Space-Time Emergence equation (3.1.3) implicitly models. To draw a parallel, this may be like trying to fathom human thought by an observation of changes in facial expression. We may erect a science of how changes in the curling of the lip, creasing of the forehead, compression of the eyes, and so on, can be linked to all manner of thought. But this is a chimera. Without considering the complexity in levels and structures and processes of thought itself, it is unlikely that facially-based surface phenomena will explain thought.

Yet, this appears to be the process being followed by contemporary quantum computing pioneers in delving into the mysteries of quantum computation.

Alternative Quantum Computing Architectures

Hence, a different kind of quantum computational process seems inevitable. Given that the quantum-level as proposed in this book is fundamentally creative, computation needs to be rethought of, as being an act of creativity leveraging the process of quantization. It is not possible for any solely physical or even quantum-physical computer, without the involvement of meta-levels, to activate such process of quantization. Further examples of such computation involving meta-levels and quantization are elaborated in a couple of the authors' pervious books on the creation of history (Malik, 2017e) and the creation of super-matter (Malik, 2018a).

If there is to be a leap in computation, it is not along a linear trajectory extrapolating what digital computers already do so that more information can be processed even infinitely faster, but rather in a manner that breaks established symmetries and creates meaningful material difference by virtue of activating the power of Light at meta-levels. But this will necessitate the involvement of instrumentation capable of penetrating meta-levels.

Following the arguments in this book, fourfold-organizations as emergences from deeper properties of Light could be such an instrumentation. The question though is when do such fourfold organizations become conscious in a way that the automatic dynamics of circumstance can be altered?

The Time-Matrix (lower left matrix) in the Light-Space-Time Emergence equation (3.1.3) reproduced below for convenience suggests that automaticity can be altered with the advent of the human, and the Space-Matrix (top right matrix) suggests that higher meta-levels can be activated when habitual patterns are broken. Hence, the human who can break habitual patterns will in general be the first instance of a fourfold-organization capable of consciously activating the meta-levels to so bring about a process of creative quantization to alter circumstance.

$$Emergence_{light-space-time} =$$

$$
\begin{vmatrix}
\begin{bmatrix}
C_\infty : [Pr, Po, K, H] \\
\left(\downarrow R_{C_K} = f(R_{C_\infty}) \right) \\
C_K : [S_{Pr}, S_{Po}, S_K, S_H] \\
\left(\downarrow R_{C_N} = f(R_{C_K}) \right) \\
C_N : f(S_{Pr} \times S_{Po} \times S_K \times S_H) \\
\left(\downarrow R_{C_U} = f(R_{C_N}) \right) \\
C_U : [P, V, M, C]
\end{bmatrix}_{Light}
&
\begin{bmatrix}
M_3 \rightarrow System_X \\
(\uparrow F \rightarrow I) \\
M_2 \rightarrow S_{System_X} \\
(\uparrow Sig \rightarrow F) \\
M_1 \rightarrow Sig_X \\
(\uparrow > P_x) \\
U \rightarrow x_U
\end{bmatrix}_{Space} \\[2em]
\begin{bmatrix}
M_3 : -\infty \leq t \leq \infty \\
\downarrow \\
M_2 : 0 \geq t > \infty \\
\downarrow \\
M_1 : 0 > t > \infty \\
\downarrow \\
U \rightarrow \begin{array}{l} t \leq E_{Cell}; TC : M_3 \rightarrow U \\ t \sim E_{Human}; TC : U \rightarrow M_3 \end{array}
\end{bmatrix}_{Time}
&
\begin{array}{l} \langle x_U \mid x_T \rangle \\[3em] TC \rightarrow x_T \end{array}
\end{vmatrix}
$$

Further, one such way to alter circumstance, or to computationally co-create may be to leverage the Nurturing-based operators specified by (3.7.14) reproduced below for convenience:

$$Nurturing\_based_{Mathematical\_operators} \ni [Remember, Linking, Relate...]$$

Specifically, leveraging (3.7.17), the Relate equation, circumstance could be consciously offered to the intelligence in a point, for that intelligence to intervene in the circumstance, for instance:

$$
Relate: \; Offer\left(x_U, \; \bigcup \begin{bmatrix} System_{Pr} \\ System_P \\ System_K \\ System_N \end{bmatrix} \right)
$$

This will likely result in space-time-energy-gravity quantization and precipitation of an enhanced logic-ecosystem in the material-fabric.

Similarly other mathematical operators could be leveraged thereby causing computation to be fundamentally different. But these can only be leveraged through the agency of deeper elements such as will, offering, love, that are the province of human-beings due to humans having developed to evolve these possibilities of Light.

Further, the progress in such creative computation will involve recording or mirroring in some kind of stratum that is architected to reflect the deep fourfold reality emergent from Light. This stratum may be a living cell that implicitly is structured by four molecular plans. The vast range of nucleic acids and proteins adds another indicative layer of subtlety that can fine-tune precision and diversification in an act of mirroring. Clearly there would have to be a period of learning preceding the human-quantum-computation is which the fourfold structure of the mirroring cells and tuned to the fourfold structure of the human operators.

Such mirroring could also leverage the process of statistical disaggregation referred to in Chapter 1.3. Hence, ecosystem-logic in the material-fabric may materialize as a range of vibrations, or waves, or influences, and sensors should be able to form a picture of the distribution of such ecosystem-logic-influences to indicate the scattering of possibility and the strongest possibility that the presiding-function is materializing as.

Through a process of experimentation, communication with such fourfold cellular structures can become more conscious, their receptivity to the balance of four-foldness that in essence is behind any act of creativity more precise, therefore allowing the link between human envisioning, activation of deeper layers, and indication of results on cellular interfaces to be completed.

Further, an amplification of the "solution" or newer ecosystem logic may be attained through a network of fourfold-organizations, be they cells, or arrangements of molecules or atoms in certain crystalline structures, distributed in the system that is the subject of such computation, to thereby achieve a broadcasting of the creative act.

Possible Application Areas

In the mathematical model of Light presented in this book all emergences are a result of the underlying fourfold properties of Light. Everything can be understood as a precise application of (3.1.3) – meaning no matter what has to be modeled, from the small to the large, it can be modeled using (3.1.3). This is not unlike using binary representation of ones and zeros to code anything. The coding scheme here is a multi-layered fourfold symmetry, capable of modeling infinite diversity, as detailed in Section 2, that captures functional to practical aspects that define any phenomenon or object. Looking back at Sections 4 through 8 of this book gives a practical sense for this coding, and similarly a sense for how anything not already illustrated in these sections, may practically be coded using (3.1.3).

This scheme of coding lends itself to phenomena such as creation and emergence, and hence to a vast range of potential creative computation applications.

Imagine corporations, or markets, that are out of synch because there is a focus on less than four of the needed organizational principles. An application of quantum-computation may be to have that corporation get back into balance, or for the balance to be created through a conscious 'willing' of sorts.

Or imagine Climate Change rushing to the point of no return, of a financial breakdown in the works threatening to throw economies into disarray. A set of human actors would be the operators to set in motion a quantum computation to set right the course of things.

Or imagine a person trying to compose music. The completeness could be realized by understanding the signature, the primary and secondary elements required to complete it, and then through an act of "seeing" in a certain way the precipitation of the needed ecosystem logic through a detailed space-time-energy-gravity quantization initiated to change the material-fabric.

These are only some examples, but the possibilities are endless.

Such a difference between construction, the focus of digital computing, and creation, the possible focus of quantum computing as elaborated in this book, is perhaps best captured by this image suggested by Albert Einstein: "Nature shows us only the tail of the lion. But there is no doubt in my mind that the lion belongs with it even if he cannot reveal himself to the eye all at once because of his huge dimension."

REFERENCES

1. Andersen, H. 1837. The Emperor's New Clothes. Denmark: C.A. Reitzel

2. Arabatzis, T. 2006. Representing Electrons: A Biological Approach to Theoretical Entities. University of Chicago. Chicago

3. Arkani-Hamed, Nima; Bourjaily, Jacob L.; Cachazo, Freddy; Goncharov, Alexander B.; Postnikov, Alexander; Trnka, Jaroslav (2012). "Scattering Amplitudes and the Positive Grassmannian". arXiv:1212.5605

4. Briggs, J. 1992. Fractals: The Patterns of Chaos. Simon & Schuster: New York.

5. Brooks, M. 2017. The Quantum Astrologer's Handbook. London: Scribe Publications.

6. Chown, M. 1990. Can Photons Travel Faster Than Light? New Scientist 126(1711)

7. Cottingham, W & Greenwood, D. 2007. An Introduction to the Standard Model of Particle Physics. Cambridge University Press. Cambridge

8. Deep Order Mathematics Videos. 2016. Deep Order Technologies. http://www.deepordertechnologies.com/new-index

9. Diamond, J. 2005. Collapse: How Societies Choose to Fail or Succeed. Viking Books: New York

10. Einstein, A. 1995. Relativity: The Special and General Theory. New York: Broadway Books.

11. Feynman, RP. 1985. QED The Strange Theory of Light and Matter. New Jersey: Princeton University Press

12. Frost, R. 2005. Elliott Wave Principle: Key to Market Behavior. New Classics Library: Georgia

13. Fuller, B. 1982. Synergetics: Explorations in the Geometry of Thinking. MacMillan Publishing Co.: New York

14. Goodsell, David. 2010. The Machinery of Life. New York: Springer

15. Gottlieb, M. 2013. The Feynman Lectures on Physics: III. California Institute of Technology. http://www.feynmanlectures.caltech.edu/III_16.html

16. Gray, T. 2009. The Elements: A Visual Exploration of Every Known Atom in the Universe. Black Dog & Levental Publishers. New York.

17. Gubser, S. 2010. The Little Book of String Theory. Princeton University Press

18. Harvard-Edu. 2014. http://news.harvard.edu/gazette/story/2014/04/harvard-to-sign-on-to-united-nations-supported-principles-for-responsible-investment/

19. Hawking, Stephen. 1988. A Brief History of Time. New York: Bantam Books

20. Heiserman, D. 1991. Exploring Chemical Elements and their Compounds. McGraw-Hill. New York.

21. Holland, P. 1995. The Quantum Theory of Motion: An Account of the de Broglie-Bohm Causal Interpretation of Quantum Mechanics. Cambridge: Cambridge University Press.

22. Hope, Chris; Fowler, Stephen J. (2007). "A Critical Review of Sustainable Business Indices and Their Impact". Journal of Business Ethics. Vol. 76. S. 243–252. Springer: New York

23. Hoverstadt, P. 2008. The Fractal Organization: Creating Sustainable Organizations with the Viable System Model. Chichester: John Wiley & Sons

24. Hyperphysics. 2016. Department of Physics and Astronomy. Georgia State University. http://hyperphysics.phy-astr.gsu.edu/hbase/mod3.html#c1

25. Isaacson, W. 2008. Einstein: His Life and Universe. Simon and Schuster. New York.

26. Jeans, J. 1932. The Mysterious Universe. Cambridge University Press.

27. Jepsen, K. 2013. Everything is Made of Fields. Symmetry. Joint Publication by Fermilab/SLAC. http://www.symmetrymagazine.org/article/july-2013/real-talk-everything-is-made-of-fields

28. Johnson, S. 2010. Where Good Ideas Come From: The Natural History of Innovation. Riverhead Books: New York

29. Kaufmann, S. 1995. At Home in the Universe. New York: Oxford University Press.

30. Kaufmann, S. 2003. The Adjacent Possible, Edge. https://www.edge.org/conversation/the-adjacent-possible

31. Knuth, D. 1968. The Art of Computer Programming. Boston: Addison-Wesley.

32. Laszlo, E. 2014. The Self-Actualizing Cosmos: The Akasha Revolution in Science and Human Consciousness. Inner Traditions: Vermont

33. Lloyd, S. 2007. Programming the Universe: A Quantum Computer Scientist Takes On the Cosmos. New York: Vintage

34. Logue, A. 2008. Socially Responsible Investing for Dummies. Wiley Publishing: New Jersey

35. Lorentz, H.A. 1925. The Science of Nature. Vol. 25, p 1008. Springer

36. Malik, P. 2009. Connecting Inner Power with Global Change: The Fractal Ladder. New Delhi: Sage Publications

37. Malik, P. 2015. The Fractal Organization. New Delhi: Sage

38. Malik, P. 2017a. Doctoral Thesis. University of Pretoria Graduate School of Technology Management. https://repository.up.ac.za/handle/2263/62779?show=full

39. Malik, P. 2017b. A Story of Light. Amazon Kindle.

40. Malik, P. 2017c. Oceans of Innovation. Amazon Kindle.

41. Malik, P. 2017d. Emergence. Amazon Kindle.

42. Malik, P. 2017e. Quantum Certainty. Amazon Kindle.

43. Malik, P. 2018a. Super-Matter. Amazon Kindle.

44. Malik, P. 2018b. Cosmology of Light. Amazon Kindle.

45. Malik, P, Pretorius, L. 2018. Symmetries of Light and Emergence of Matter. Indian Journal of Science & Technology. http://www.indjst.org/index.php/indjst/article/view/110789.

46. Mandelbrot, B, Hudson, R. 2006. The Misbehavior of Markets: A Fractal View of Risk, Ruin, and Reward. New York: Basic Books

47. Mora, C, Tittensor, D, Adl, S, Simpson, A, Worm, B. 2011. How Many Species Are There on Earth and in the Ocean? Plos.org. http://journals.plos.org/plosbiology/article?id=10.1371/journal.pbio.10 01127#abstract0

48. Narby, J. 1998. The Cosmic Serpent: DNA and the Origins of Knowledge. New York: Penguin Putnam.

49. NASA-darkmatter. 2016. http://science.nasa.gov/astrophysics/focus-areas/what-is-dark-energy/

50. Ogburn, W. Tomas, D. 1922. Are Inventions Inevitable? A Note on Social Evolution. Political Science Quarterly 37, no.1

51. Olive, K.A et al. 2014. Particle Data Group. Chin. Phys. C, 38, 090001.

52. Openshaw, J. 2015. http://www.marketwatch.com/story/socially-responsible-investing-has-beaten-the-sp-500-for-decades-2015-05-21

53. Particle Data Group. 2015. Lawrence Berkeley National Laboratory. http://www.cpepphysics.org/main_universe/universe.html

54. Patterson, K, Grenny, J. Crucial Conversations: Tools for Talking when Stakes are High. 2011. McGraw-Hill: New York.

55. Pauli, W. 1964. Nobel Lectures, Physics 1942 – 1962. Elsevier Publishing Company. Amsterdam.

56. Pearson, R. 1997. "Frontier Perspectives", Journal of the Center for Frontier Sciences at Temple University. Spring/Summer 1997, Volume 6, Number 2, ISBN: 1062-4767.

57. Perkowitz, S. 2011. Slow Light. London: Imperial College Press
58. Planck, M. 1933. Where is Science Going? Ox Bow Press. Connecticut.
59. Portugali, J., 2012. *Self-organization and the city*. New York: Springer Science & Business Media.
60. Prigogine, I. 1977. Time, Structure, and Fluctuations. *Nobelprize.org*. Nobel Media AB 2014. Web. 5 Mar 2016. <http://www.nobelprize.org/nobel_prizes/chemistry/laureates/1977/prigogine-lecture.html>
61. Rovelli, C. Reality Is Not What It Seems. New York: Riverhead Books
62. Salkind, N. 2007. Encyclopedia of Measurement and Statistics. Thousand Oaks: Sage Publications.
63. Snyder, M. 2010. Stanford Medicine. http://med.stanford.edu/news/all-news/2010/03/what-makes-you-unique-not-genes-so-much-as-surrounding-sequences-says-stanford-study.html#.html
64. Sri Aurobindo. 1971. Social and Political Thought. Sri Aurobindo Ashram Press: Pondicherry
65. Stewart, Ian. 2012. In Pursuit of the Unknown. Basic Books. New York.
66. Swimme, B. 2001. The Universe is a Green Dragon. Bear & Company: Rochester
67. Smith, J, Szathmary, E. 1995. The Major Transitions in Evolution. Oxford: Oxford University Press
68. Toynbee, A. 1961. A Study of History, Volumes I – XII. Oxford University Press: Oxford
69. Turok, N. 2012. The Universe Within. Anasi Press: Ontario
70. Tweed, M. 2003. Essential Elements: Atoms, Quarks, and the Periodic Table. Walker & Copmany. New York.
71. Weizenbaum, J. 1976. Computer Power and Human Reason: From Judgment to Calculation. San Francisco: W.H. Freeman
72. Wheeler, J. Ford, K. 2000. Geons, Black Holes, and Quantum Foam: A Life in Physics. New York: W. W. Norton & Co.
73. Whitaker, A. 2006. Einstein, Bohr and the Quantum Dilemma: From Quantum Theory to Quantum Information. Cambridge: Cambridge University Press
74. Wilczek, F. 2016. A Beautiful Question: Finding Nature's Deep Design. New York: Penguin Books
75. Willis, A. 2003. The Role of the Global Reporting Initiative's

Sustainability Reporting Guidelines in the Social Screening of Investment. Journal of Business Ethics. Volume 43. Springer: New York

76. Wolchover, N. 2013. A Jewel at the Heart of Quantum Physics. Quanta Magazine . https://www.quantamagazine.org/20130917-a-jewel-at-the-heart-of-quantum-physics/

77. Wright, R. 2009. Evolution of Compassion. https://www.ted.com/talks/robert_wright_the_evolution_of_compassion/transcript?language=en. TED 2009.

78. Yates, F.E. 2012. *Self-organizing systems: The emergence of order.* New York: Springer Science & Business Media.

Author's Previous Books

Early Books

1. The Flowering of Management
2. India's Contribution to Management

Fractal Series

1. Connecting Inner Power with Global Change: The Fractal Ladder
2. Redesigning the Stock Market: A Fractal Approach
3. The Fractal Organization: Creating Enterprises of Tomorrow

Cosmology of Light Series

1. A Story of Light: A Simple Exploration of the Creation and Dynamics of this Universe and Others
2. Oceans of Innovation: The Mathematical Heart of Complex Systems
3. Emergence: A Mathematical Journey from the Big Bang to Sustainable Global Civilization
4. Quantum Certainty: A Mathematics of Natural and Sustainable Human History
5. Super-Matter: Functional Richness in an Expanding Universe
6. Cosmology of Light: A Mathematical Integration of Matter, Life, History & Civilization, Universe, and Self